I0480111

Chemistry & Materials

for

Civilization

George E. Parris

Copyright 2020

Preface

Civilizations have been defined by the materials that drive their economies. The Stone Age, Bronze Age and Iron Age each lasted thousands of years; but starting in the 1700's of our current era, knowledge has expanded so fast that new inventions in chemistry and materials have come almost yearly and the applications of materials in fields such as construction, manufacturing, energy, transportation, communications, computing and medicine have produced an economy that is not readily characterized in a single term. Within the lifetimes of people whom I have personally known (my grandparents and parents) we have advanced from an economy dependent on horses to an economy that includes space travel, molecular biology, and the search for extra-terrestrial life.

Meanwhile, in my chosen profession of chemistry, I feel that very little has changed in our fundamental knowledge base in my lifetime. The topics that I teach to undergraduate students in 2020 are essentially the same that I learned in 1970. I was fortunate to be educated after an understanding of the electronic configuration of the atom was in place and techniques including spectroscopy, crystallography and chromatography were well established (if not fully developed).

Perhaps the greatest change in chemistry in the last 50 years has been in how it is perceived. I was attracted to chemistry because in 1950 it was seen as the science producing materials that improved our daily lives. I believe the phrase was "better living through chemistry" (Dupont 1935-1982). Nonetheless, by the time I graduated from Georgia Tech with a PhD in 1974, it was recognized that there were problems. Mass production of useful products was unwittingly causing problems with pollution of local and global significance. In fact, this was a field that I thought would be interesting to pursue: Environmental chemistry. But literally with my first job with the US Environmental Protection Agency, I discovered that the topic of environmental chemistry (unlike laboratory research) was very closely comingled with politics and profits. What I had assumed would be a rational scientific endeavor was (and is) corrupted by the processes of seeking political power over and corporate profits from a general population with whatever tools can be brought into effect.

Moreover, I see "chemistry" as a science becoming rather stagnant and limited. Little has changed in the undergraduate education in 50 years. And some important parts (thermodynamics and quantum mechanics) are actually taught with less rigor. Meanwhile, entirely new fields *that depend upon chemistry* are growing entirely outside the academic sphere of chemistry. I'm thinking primarily about "material science and engineering" and "molecular biology." Academic chemists ignored these fields. It is remarkable that *the most important application of structural organic chemistry* was actually the work of a biologist and a physicist (Watson and Crick, 1953). I am not in a position to save academic chemistry from what appears to be a state of stagnation (if not retrenchment), but I would like to arm the general public with an appreciation of what chemistry has and can accomplish.

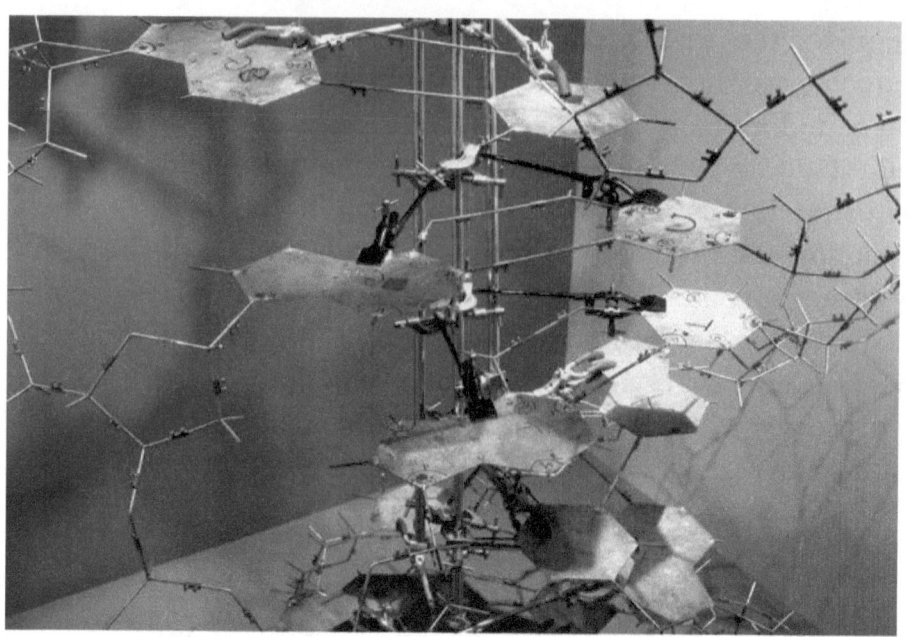

Watson and Crick's DNA Molecular Model from 1953

Image Credit: Science Museum (London)

Over the last ten years, I have taught courses primarily to students who are not going to spend their lives studying chemistry and who in some cases have little interest in science or math at all. Nonetheless, they live, work, pay taxes and vote in a world where chemistry is important. Thus, I have written this book to convey what I what I think the members of the general public (including professional chemists) need to know about chemistry and its role in modern society and the advancement of civilization. Basically, what a member of the public should know to make informed decisions in their personal, occupational and political activities. Indeed, as much as professional chemists are part of the general society, much of the material is needed by them (but not taught in any chemistry major course I am aware of).

The book is written in three parts:

Part I provides a general background in chemistry. It is similar to the material covered in general chemistry courses, without the math. Obviously, a chemist or chemical engineer who has mastered this and more, could skip Part I, but would likely find the remainder of the book more useful than other professions.

Part II focuses on the interaction of chemical technologies with advancement of human culture and society. The approach is strongly from a historical perspective showing how and why chemistry and society advanced together. In particular, I am interested in business, commerce and politics as well as human health through the ages.

Part III moves specifically into the realm of biologically active molecules that have a special place in that they go beyond being "materials" used to make things and actually directly impact health and wellbeing of humans. These, of course, fall into categories such as food additives, drugs, pesticides, chemical weapons, and "environmental pollutants," which because of their direct impact on human health have prompted the introduction of a system of regulations, which (in turn) directly impacts innovation and commerce.

Part I. Introduction

1.0 The Physics of Chemistry

1.1 Conservation of Energy in its Many Forms

Perhaps the most important and fundamental law of physics is that "energy is conserved." This, law applies to *chemical reactions* (which do not involve and changes of mass) and nuclear processes (e.g., radioactivity) that typically do involve changes in mass. Chemical processes (as wee will discuss below) involve changes in the electron fields that act as "glue" holding atoms together. On the other hand, radioactivity involves changes in the nuclei of atoms these changes typically change the mass of the system and the mass is converted into some sort of energy (e.g., potential energy, kinetic energy or electromagnetic radiation, such as gamma rays). Albert Einstein's famous equation form 1906 provided the conversion factor for energy and mass[1]:

---

[1] **E** stands for energy, **m** stands for mass, **C** is the speed of light: a constant equal to 3.00×10^8 m/s and the subscripts **i** and **f** stand for initial and final. Note that when $m_f < m_i$, the energy change (ΔE) has a negative sign indication that the system is releases energy (i.e., we use the term "exothermal").

$$E = mC^2$$

I actually prefer to express this equation as

$$(E_f - E_i) = (m_f - m_i)\ C^2$$

$$\Delta E = \Delta m\ C^2$$

This form indicates that a change in mass (Δm) must be associated with a change in energy.

It probably goes without saying, but I want to emphasize that not only is energy conserved, *"energy is conserved instantaneously."* In addition, while energy must be conserved, it may be conserved in any of its forms: potential (including chemical), kinetic, electromagnetic and mass.

We will discuss electromagnetic radiation in later sections, but I will mention here that Max Planck (1858-1947) was the person who associated the energy of electromagnetic radiation (i.e., the photon) with the properties of waves:

$$E_{photon} = h\nu = hc\ (1/\lambda)$$

Where "h" is Planck's constant (6.6×10^{-34} J.s/photon), "ν" is the frequency of the wave, "c" is the speed of light and "λ" is the wavelength.

1.2 Systems Tend towards their Lowest Energy State

You are certainly familiar that fact that a ball spontaneously rolls down hill and requires effort to make it go up hill. This is a fundamental concept of physics and chemistry. We can define a system any way we want to, but within that system changes will tend to spontaneously occur that will minimize the energy content of that system: The system will tend to move to a relatively more stable state. However, there may be an energy barrier that the system cannot overcome, which prevents it from reaching the lowest available energy state.

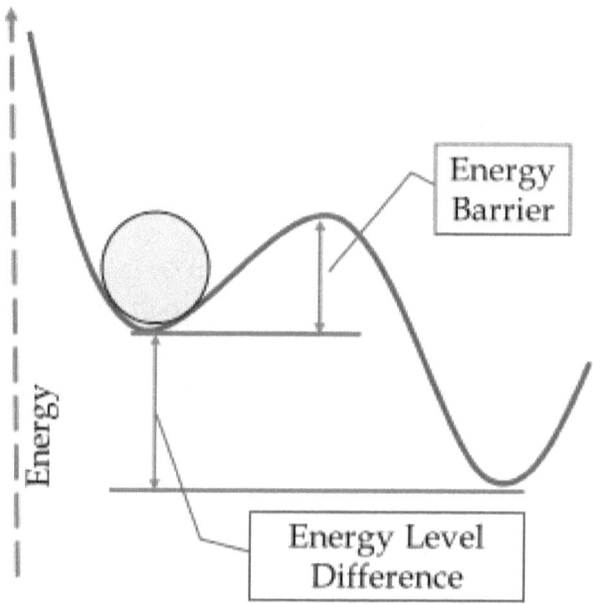

That Energy Barrier may be large or small. Chemists call the barrier, "activation energy." The activation energy (Ea) determines the rate of the reaction, but it does not affect the ultimate direction of the reaction. The direction of the reaction is determined by relative energy differences of the starting condition (e.g., reactants) and the ending condition (e.g., products).

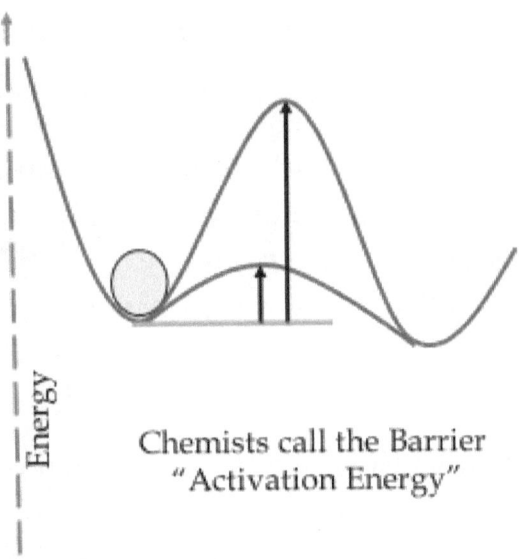

Chemists call the Barrier
"Activation Energy"

The "rolling down-hill" model is useful, but can be mis-leading for chemists.

Heat

The most notable change during a chemical reaction is often the change in temperature of the system. *Heat* is the kinetic energy (i.e., energy associated with movement) of atoms and molecules as they translate, rotate and vibrate. Heat seems to flow through a medium and disperse. This effect is the result of collisions of molecules and atoms in such a way that momentum is transferred from one particle to another. Molecules with high kinetic energy are more likely to lose energy in a collision than molecules with low kinetic energy.

Momentum can be rotational momentum or linear momentum but the total momentum is always conserved in collisions. On the other hand, simple calculation show that kinetic energy cannot be conserved when momentum is conserved. Some kinetic energy is always lost, which means that some of the kinetic energy must be converted to other types of energy. In particular, some energy may be converted to potential energy of chemical bonds; and ultimately, any energy not otherwise conserved as kinetic or potential energy is released as a photon of energy (electromagnetic radiation) which we will discuss below.

It is difficult to measure heat of a system directly; however, the temperature (measured typically by the expansion of a fluid in a thermometer) of the system is directly proportional to the heat (kinetic energy) of the medium. The proportionality constant is called the *heat capacity* (Cv) of the system[2]:

Heat = Cv x Temperature

Overall, the movement of heat is called *thermodynamics*, and chemists must be aware that energy can take on many forms. Heat (i.e., the kinetic energy stored in atoms and molecules) *per se is not conserved in chemical reactions, energy, which may be in many forms, is conserved.* And systems tend to move to lower

---

[2] Cv indicates heat capacity at constant volume. If the volume is allowed to change, the system may do work (e.g., like the movement of a piston) which requires energy.

energy, thereby dispersing energy (not necessarily heat) to the *surroundings* of the system.[3] Ultimately, chemists have developed a system for accounting for "energy" (G) where "heat" is very carefully defined as "enthalpy" (H) and the other energy elements are defined as an effective heat capacity "entropy" (S) at the temperature of interest (T).

$$G = H - TS$$

Note that the temperature scale that it used in thermodynamics is the scale where zero (i.e., *absolute zero*) is set at the point where *all heat that can be withdrawn from the system to do work has been withdrawn* ($0^{\circ}K = -273^{\circ}C$). (See for example: Parris, Thermodynamics 3rd edition, 2019).

1.3 Electrostatic Attraction

Ion-Ion Interactions

In ancient times, people realized that rubbing various materials together caused a situation in which the materials would repel or attract small particles. For example, running a comb through long, dry hair may make the strands of hair repel one another.

---

[3] We still retain the terms "endothermic" and "exothermic" to refer (respectively) to energy entering or leaving a defined system. But these days, we understand that reactions can occur spontaneously (with dispersion of energy) although heat may be flowing into the system.

Amber was particularly associated with the phenomenon; and the name for this property of amber (*electricus*) entered the English language in the 1600s as "electricity." By the 1700s, this phenomenon was associated with sparks of light and weak forces among "light bodies."

It was not until 1785 that Charles Augustin de Coulomb invented a way to measure such small forces. His invention was the torsion balance in which a long thin metal wire acted as a spring in which the twisting deflection was proportional to the force applied.

Torsion Balance (circa 1785)

In the photograph at the left, the wire extends down the tall tube to the lower chamber where a balanced cross-arm is attached. On one end of the cross arm a small sphere of nonconducting material is affixed. An identical sphere is attached to a rigid post mounted in the lower chamber. The entire device is protected from air currents and is on a leveling base. The deflection of the spheres is measured by an angular scale on the outside.

The rigid post can be easily removed and charged with static electricity by rubbing. Coulomb reasoned that since the spheres were identical, when they were contacted, half of the charge went to the movable sphere. He could then measure the deflection. By removing the post and discharging the rigid sphere he could then reassemble the apparatus and transfer half of the charge from the mobile sphere to the rigid sphere. In this way he could develop a mathematical relationship between the charge and the deflection (i.e., distance between the centers of the spheres).[4] The resulting law is an inverse square law for electrostatic force (analogues to the gravitational law):

$$\textbf{Force} = \textsf{K}\ \textbf{q}_i\ \textbf{q}_j / \textbf{r}_{ij}^2$$

In this equation, **q** stands for the charge (which can be + or -) and **r** stands for the distance between the centers of charged spherical particles. Note that a (+) x (-) results in a negative force (attraction) and a (+) x (+) or a (-) x (-) results in a positive force (repulsion). Obviously, when r is very large (i.e., approaches infinity) the force between the particles (i and j) goes to zero. Thus, we set infinite separation as the condition of zero potential energy and convert the equation to measuring energy by multiplying by **r** as follows:

Force x distance = Energy

---

[4] There is a principle of electrostatics that states that when you have a homogeneous spherical charge, the entire charge can be assumed to be in the *center* of the sphere (Gauss's law).

Thus, we can write Coulomb's law in terms of potential energy between the two particles (i and j):

$$E_{ij} = K\, q_i\, q_j /\, r_{ij}$$

where[5]: $K = 1.39 \times 10^3$ kJ.Å/mole

Notice, that if we have more than two particles, all of the interactions (i,j) are *independent* and all we have to do is *add up the energies associated with all the pair-wise (i, j) interactions* to obtain the electrostatic potential of the system (e.g., a crystal lattice):

Energy of a system of charged particles = Σ Eij

This is a very important principle in chemistry and can be used to calculate the energy of charged particles in different relative positions. We will do this later, when we talk about the structure of atoms and molecules.

---

[5] Physicists and most chemistry books present Coulomb's constant in units that are not easy for chemist to apply. These are the equivalent units that are useful to chemists. The units are "kilo-Jules" (kJ, a unit of energy), "Angstrom" (Å = 10^{-10} m), "mole" (a counting unit equal to 6.02×10^{23} molecules). Chemists use the "mole" the way doughnut shops use the term "dozen."

Page 14 of 492

Ion-Dipole Interactions

It turns out that in nature some molecules (even those with no net charge) have enough "charge separation" to exhibit electrostatic behavior. Consider a case where a particle (with no net charge) has a partial separation of charge that can be represented as

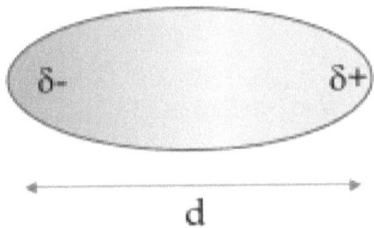

A dipole has partial charge separation
and
The *dipole moment* (μ) is calculated as
the charge times the distance of separation

$$\mu = \delta d$$

In an ion-dipole interaction, one end of the dipole is always attracted by the ion, so the net interaction is always attractive. Moreover, we know that as the separation between the ion and dipole (r) become very large with respect to the separation of the partial charges in the dipole (d):

$$r >> d$$

The dipole is essentially neutral molecule. Thus, the ion-dipole potential can be represented as an inverse-square law:

$$E_{ij} = - K \, |q_i| \, \mu_j / r_{ij}^2$$

Which is based on Coulombs law.

1.4 Electromagnetic Radiation

The speed of Light

Of course, humans have recognized light and dark for all of our existence and light has been associated with the sun (our star), but it was not until the Renaissance that people in Europe accepted that the moon was illuminated by the sun and we began understanding the heliocentric arrangement of our solar system. Naturally, questions were raised about the nature of light. Gunpowder (a.k.a., black powder) was introduced into Europe about 1300. This technology soon made it apparent that light (a flash and puff of smoke) reached the observer before sound (the bang). The difference is barely noticeable at 100 m but with loud explosions, the delay of the sound is clearly apparent at 1000 m. This set Galileo Galilei (1564-1642) wondering about the speed of light and he is credited with attempting to measure the speed of light using shutter lanterns sending flashes of light back and forth among surveyed observation post. Light was simply too fast to measure with that method, but it allowed a reasonably good velocity of sound to be determined (330 m/s).

Nonetheless, with his telescope Galileo discovered three moons of Jupiter (1610). The orbits of these moons provided something of a celestial clock that could be observed (weather permitting) most places on earth. Thus, they were soon applied to the problem of estimating longitude by sailors, who were now making voyages around the world and attempting to map new continents. There were a number of problems with the technique (especially on a rolling ship in the open sea). But one issue was particularly interesting.

Copernicus (1473–1543) had developed the idea of a heliocentric solar system with planets orbiting around sun. It was clear that Jupiter was farther from the sun than the earth and sometimes the earth was away from Jupiter approaching it a great speed and other times the earth was near to Jupiter moving away from it at great speed. These differences in relative speed did not seem to matter[6], but Ole Roemer (1644–1710) noticed that the

---

[6] When the earth was at points in its orbit that would be the same distance to Jupiter the same orbital period was observed, whether the earth was approaching or receding from Jupiter. This is a potentially important point, but was within the experimental error of the day.

distance of the earth from Jupiter did make a difference in the apparent orbital period of the moon Io (1.769 Earth days). The important point was that when the earth was nearest Jupiter the eclipses of Io by Jupiter was 11 minutes shorter than average and when the earth was farthest from Jupiter the eclipse was 11 minutes longer than average. Io was a like flashing lantern...a very long way away. Romer announced the variation in the period of eclipse and its reason on August 22, 1676. But the actual calculation of the speed of light (about 2.1×10^8 m/s) was apparently done by Christiaan Huygens (1629–1695) using elementary assumptions about the diameter of the earth's orbit. The important point was that the speed of light was not infinite and an order of magnitude value was obtained.

Light: Particle or Wave?

Romer and Huygens were Danish colleagues. But, in Britain Isaac Newton (1642–1726) was conducting experiments on optics. Newton's interpretation of light was that it was a particle of some sort. Huygens (perhaps because of the apparent insensitivity of the measured speed of light to the relative motion of the earth and Jupiter) was inclined to believe that light was some sort of wave (like ripples on a pond). With by far the larger reputation, Newton's idea won public acceptance throughout the 1700s.

Then along came Thomas Young (1773–1829). Young was a brilliant polymath who was a keen observer with an open mind.

He noticed that if he held a card slightly less than 1 mm thick (i.e., 0.85 mm thick) in a collimated beam of light, it produced fringes of different colors on a screen, and if light was blocked on one side of the card, the fringes disappeared. This behavior could only be explained by light as a wave of some sort.

In a more elaborate experiment, light was passed through a single slit and then through parallel slits to produce two synchronized sources of light. He observed that the pattern of the light on a screen was a series of parallel lines with shadows in between. This experiment could only be interpreted as interference (i.e., cancellation and reinforcement) of the synchronized waves. Importantly, by knowing the (i) separation of these lines, (ii) the separation of the slits and (iii) the distance from the slits to the screen, the wavelength (λ) could be calculated from geometry.

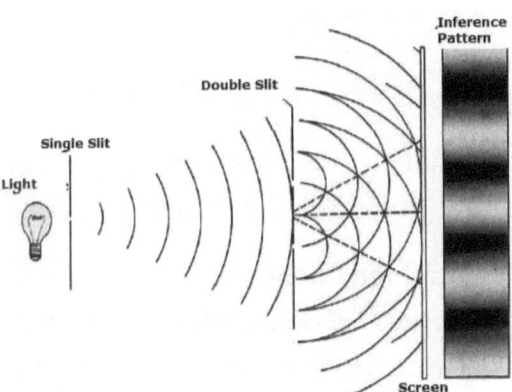

Source: QUORA: https://www.quora.com/In-a-Michelson-interferometer-how-do-fringes-form

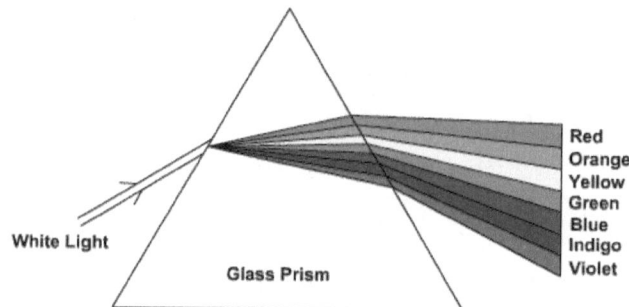

Author: Prince Singh; Source: https://www.quora.com/If-a-ray-of-light-is-passing-through-a-prism-which-colour-appears-at-the-top-of-the-spectrum

Newton had already shown that white light was composed of many colors. Thus, in 1801, the wave nature of light was firmly established and people soon learned that the different colors of light obtained by passing sunlight through a prism represented different wavelengths.

UV and IR and the Solar Spectrum

It soon became clear (that the visible) light had wavelengths between about 350 nm (blue) and 700 nm (red). Was this all?

The astronomer William Herschel (1738–1822)[7] was looking for filters to observe sun spots (1800) and discovered that filters that absorbed red light became much warmer than filters that absorbed blue light. Conversely, the blue light beam was cooler

---

[7] Father of John Herschel (1792-1871).

than the red-light beam. He followed up this observation by breaking up sunlight with a prism and allowing the various bands to fall on a series of thermometers. The farther to the red end of the spectrum he went, the warmer the thermometers became. He even tried beyond the ends of the spectrum and discovered that below red (i.e., *infrared*) there was invisible light that was even warmer than visible red.

In a similar experiment, Johann Ritter (1776–1810) compared the rate of darkening of silver chloride[8] from the red end of the spectrum to beyond the violet end. He discovered that beyond the visible violet color the ability to darken silver chloride was greatest. He called this invisible light "chemical rays," but it was soon described as *ultraviolet*.

Thus, by 1801, it was known that *visible light* was only a limited range of a wide spectrum of energy coming from the sun or visible light sources. Recall that this was the time that Thomas Young was arguing that light was a wave phenomenon with his interference through narrow slits.

The spectrum of light from the sun was measured both in its range (i.e., wavelength) and in its intensity. In 1838, Claude

---

[8] It had been known for a long time that light acted on silver compounds to produce dark colors (e.g., 1614, Angelo Sala (1576-1637)). Carl Scheele (1742–1786) managed to react chlorine gas with silver to make silver chloride (1774), which is very insoluble in water. When silver chloride first precipitates it is white, but upon exposure to light it darkens to a purplish and then a black color as it decomposes to particles of elemental silver.

Pouillet (1790 – 1868) developed a pyrheliometer to measure the total amount of energy arriving on the earth from the sun. These devices use a dark (blackbody) receptor[9], of known surface area, that presumably collects all radiant energy without bias of wave length and transfers that energy as heat to a measured volume of water. The rate of change of the temperature of the water (with a known heat capacity) is measured to determine the rate at which radiant energy falls ($j/s/m^2$ = watts/m^2) on the receptor. By using a prism or grating the energy of a specific range of wavelengths can be measured and it is customary to plot the results in watts/m^2 per nm of wavelength (see below).

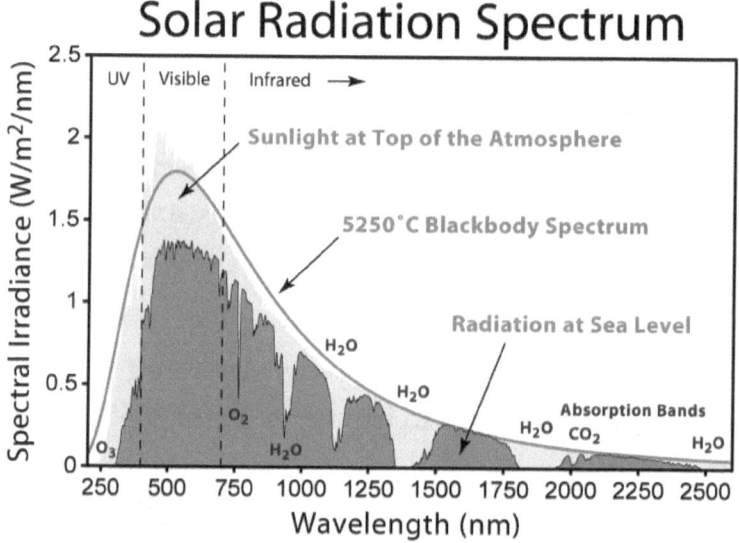

Author Robert A. Rohde, Source Wikimedia Commons.

---

[9] If you mix all colors of paint together you get black, because all the colors are absorbed.

This process has been vastly improved over the years[10] and a modern diagram is presented above.

Faraday and Maxwell

Michael Faraday (1791 –1867) began as an assistant to Humphrey Davy who was studying the ability of electricity to separate metals from their compounds. Davy's interest progressed to study of electric fields and their effects on magnetics (e.g., compass needles). Faraday continued this work and recognized the phenomenon of electromagnetic induction. Clearly, there was some mechanism for the interaction of electrostatic and magnetic forces. One of the many questions was how fast does electric charge flow through a wire. It was obviously very fast, but not measurable in the mid-1800s.

James Clerk Maxwell (1831–1879) was interested in the transmission of radio signals. These were obviously waves (based on the work of Heinrich Hertz (1857–1894)) that originated from electromagnetic phenomena in conductors and induced electromagnetic phenomena in conductors, but were they associated with light? Maxwell realized that the equations that governed electric forces (i.e., Coulomb's equation) and magnetic forces each had fundamental constants (K and B):

---

[10] 1858, Johann Müller (1809 – 1875); 1872, Sergi Lamansky (unknown).

Electric field = K q/r^2 = $q/4\pi\ \varepsilon_0\ r^2$

Magnetic field = B μ_0 I/2πr

These fields are perpendicular to one another: Current flows along a wire and produces a magnetic field radiating perpendicular to the wire.

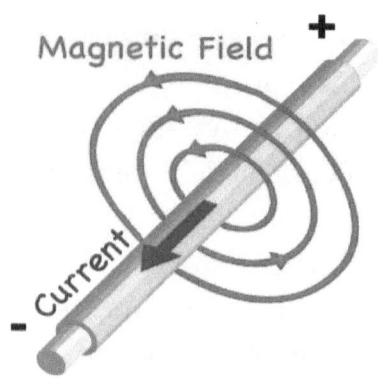

Source: https://packetpushers.net/back-basics-power/

Where ε_0 and μ_0 are properties of free space through which the electromagnetic waves travel. These constants had been determined independently and Maxwell discovered that

$$1/(\varepsilon_0 . \mu_0)^{1/2} = C \quad \textit{the speed of light (3.00 x 10}^8 \textbf{ m/s)}$$

Thus, visible light[11] was part of part of general phenomenon known as *electromagnetic radiation*, which can be represented as

---

[11] By this time, the speed of light had been determined with more precision by various experimenters.

perpendicular electric (E) and magnetic (B) vectors traveling through space carrying momentum.

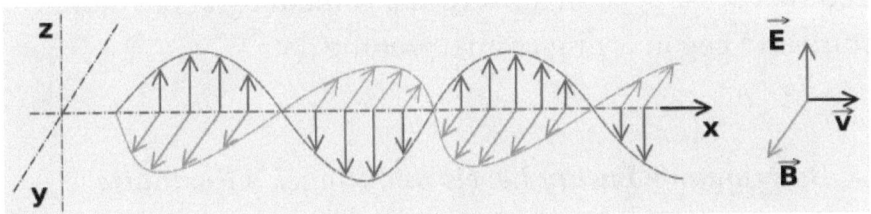

Author: SuperManu; Source: Wikimedia Commons.

Characteristic Colors of Elements in a Flame

It had been known for a long time that different elements produce different colors when placed in a flame. Robert Bunsen (1811–1899) and Gustav Kirchhoff (1824–1887) teamed up to pass light *emitted* from elements in flames through a prism and observed that the light was made up of a series of distinct lines that were unique for each element (1856-1858). This technique (known as "atomic spectroscopy") was soon applied for qualitative analyses of minerals and compounds.

Part of their research involved finding a way to produce a continuous spectrum (without absorption or emission lines). The Bunsen burner flame contained few lines but did not produce a continuous spectrum. Kirchhoff used incandescent lime (very high melting calcium oxide) to produce a continuous spectrum. With the continuous spectrum source, he could verify that the *absorption spectra* of elements matched their *emission spectra*.

Atomic spectra were being collected on every new element that could be isolated. By 1880, Johannes Rydberg (1854–1919) realized that these spectra must reveal something about atomic structure and began trying to interpret the data.

Boltzmann's Energy Levels and Planck's Postulate

Kirchhoff had opened a new field in 1859 with the study of blackbody radiation. He posed the question of how the intensity and wavelength of radiation from a surface change with temperature. Kirchhoff introduced the idea of a device called a "cavity radiator," which perfectly absorbs and emits radiation of all wavelengths. [12]

The term "blackbody" radiation was introduced by Kirchhoff in 1860 to describe a surface that would emit or absorb all wavelengths of light perfectly. Kirchhoff's revelations were just in time to be interpreted in the electromagnetic wave concept being formulated by Maxwell between 1860 and 1871. In 1879,

---

[12] Concurrently, Balfour Stewart (1828–1887) was looking at radiant transfer of energy. Of course, by this time radiant heat was recognized to be a wave phenomenon; and by examining emittance and absorbance of radiant heat, he was able to prove that the ability of a source to emit was equal to its ability to absorb. Moreover, dark colored objects (soot and charcoal) were better at both absorbance and emittance that other materials. Stewart did not publish this until 1860 so it was unknown to Kirchhoff and Bunsen who published the same idea in 1859.

Joseph Stefan (1835-1893) determined that the total radiation from a blackbody is proportional to the absolute temperature raised to the 4th power. This law was extended by his student Ludwig Boltzmann (1844-1906) in 1884.

At this point, we have not really said anything about the spectrum obtained from a blackbody other than that it is continuous. That is, in principle, every wavelength is represented in the blackbody spectrum. But anyone who has watched a blacksmith work or turned on the eye of an electric stove realizes that the most intense electromagnetic waves coming from an object move to shorter wavelengths as the temperature rises. For example, a piece of iron (or anything else) does not produce any visible glow (i.e., radiation that we can observe with our eyes) although it may be exceeding hot to the touch. Continued heating produces a dull red glow (525 °C), and then a brighter red (690°C) to orange glow (845°C), and finally a yellow (1000°C) to white (1205°C) glow. Soon thereafter, iron melts (1,538°C). In 1893, Wilhelm Wien (1864-1928) developed an empirical relationship between the *wavelength of maximum energy intensity*[13] (λ_{max}) and the surface temperature of an object that radiated at all wavelengths.

$$T \approx 2900 \text{ } \mu m \text{ }^\circ K/(\lambda max)$$

You will notice from the diagram of the solar spectrum (above) that light reaching the top of earth's atmosphere closely

---

[13] It is convenient to discuss this in terms of radiance (i.e., spectral radiance) which is defined as the power (watt) per solid angle (in steradian (sr) equivalent to squared radian) per square meter: $W/sr.m^2$.

approximates the blackbody radiation curve for a surface temperature of 5250°C (about 5520°K). But atmospheric gases (e.g., ozone, oxygen, water and carbon dioxide) somewhat diminish the energy[14] that actually falls on the surface of the earth.

In the late 1800s, measurements at relatively low temperatures (300°K) suggested that the radiance increased as the wavelength decreased. This inspired John William Strutt, (a.k.a. Lord Rayleigh) (1842–1919) and James Jeans (1877–1946) to attempt to explain the relationship through a derivation based on classical mechanics. The model they chose to analyze was a cavity inside a blackbody at thermal equilibrium. The number of modes of electromagnetic radiation that could be present in such a cavity (per unit frequency per unit volume) is given by (v = frequency, c = speed of light):

$$8\pi v^2/c^3$$

The problem with classical mechanics was that all modes were equally probably and the partitioning energy equally among all the modes resulted in a prediction that at high temperature the amount of energy in short wave lengths (high frequencies) would simply increase without limit (sometimes called the "ultraviolet catastrophe"). This model was proposed about 1900

---

[14] Some of the energy is scattered or absorbed and re-irradiated randomly into space or stored as potential energy in chemical reactions.

and was almost immediately superseded by a better and independent idea.

Ludwig Boltzmann (1844 –1906) had developed a kinetic theory of gases and interpretation of entropy, that required the existence of atoms and molecules.[15] As early as 1872, Boltzmann was considering discrete energy levels; and in 1897, he suggested the use of statistics in trying to understand blackbody radiation.

For the purposes of this discussion, Boltzmann's idea that energy in a (molecular) system would not be evenly distributed among all possible configurations, but rather would be concentrated in the lowest energy states by the need to maximize entropy. For example, in a system of 7 molecules with an energy-level spacing of 0, 1, 2, 3, 4, etc. units of energy and with a combined (system) total of 13 units of energy (see chart below).

Classical mechanics (ignoring the restriction to energy levels) would assume that each molecule had 13/7 of a unit of energy (equipartition of energy as in the Rayleigh-Jeans hypothesis). But if the molecules were constrained to only certain energy levels (or the electromagnetic waves were confined to only certain wave lengths) corresponding to transitions among the discrete energy levels of the system, then maximization of the number of microstates (highest entropy) would suggest a

_____

[15] It is interesting that while chemists had accepted the idea of atoms and molecules decades earlier, classical physicists were not convinced until the work of Einstein on Brownian motion in 1906.

different (statistical) distribution of energy among the molecules (or electromagnetic waves).

Energy Level	Micro-state 1
level 4	X $1 \times 4 = 4$
level 3	XXX $3 \times 3 = 9$
level 2	
level 1	
level 0	XXX $3 \times 0 = 0$
Total energy	13
$S/k_B = \ln W$	$\ln(5040/36) = 4.94$
$W = N!/\Pi(n_i!)$	$W = 7!/(1!)(3!)(3!)$

Note that the calculation above is only one of 5 microstates that that would maximize the system's entropy. The sum of these five microstates[16] provides a Boltzmann distribution of energy among the energy levels (i.e., most molecules or waves in the lower energy levels). (See Parris, Thermodynamics, 3rd Edition,

---

[16] There are five microstates that each have the same maximum Boltzmann entropy and would each contribute equally to the state of the system, yielding a Boltzmann distribution of (7 x 5 =) 35 elements (1 + 2 + 3 + 5 + 7 + 8 + 9 = 35) spread among 7 energy levels (6/5/4/3/2/1/0).

2019 paperback, 2015 digital for a full explanation of Boltzmann entropy.

In the case of electromagnetic waves, the maximum radiance (i.e., the highest energy density) would tend to fall towards the middle of the spectrum (as shown in this qualitative example):

Energy Level	0	1	2	3	4	5	6
Low Temperature Occupancy (Boltzmann Distribution)	50	20	10	5	2	1	0
energy x occupation (implied radiance)	0	20	20	15	8	5	0

Energy Level	0	1	2	3	4	5	6
High Temperature Occupancy (Boltzmann Distribution)	2	8	17	25	30	2	1
energy x occupation (implied radiance)	0	8	34	75	120	20	6

If you accept this, then all you need to know is what the relationship between energy and wavelength (or energy and frequency) is.

Max Planck (1858-1947) published his *Treatise on Thermodynamics* (1897) and was aware of Boltzmann's ideas. It did not take him long to postulate that the energy carried by electromagnetic radiation must be related to the wavelength (and frequency) and all he had to do was find the proportionality constant (h):

$$E_{photon} = h(c/\lambda) = h\nu$$

Planck essentially started with the same analysis of waves in a blackbody cavity that Rayleigh and Jeans had used and assigned energies to those waves and coupled the electromagnetic energies to a distribution function that had been developed by Boltzmann.[17]

It should be noted that Planck, who was devoted to classical physics, came to this conclusion with great trepidation. Boltzmann's ideas were not widely accepted by physicists.

---

[17] As expected, the Wien relationship can be derived from the Planck equation at high temperatures and the results from the Planck equation and the Rayleigh-Jeans equations converge at low temperatures. All of this happened almost simultaneously (1900) with the Rayleigh-Jeans work, which was immediately superseded by Planck's blackbody radiation equation.

Indeed, Boltzmann met so much criticism that he became depressed and committed suicide in September 1906. There was no classical physics that could be used to deduce Planck's relationship. Planck presented it as a "postulate" (i.e., an unsupported assumption that seems to work) and waited for criticism to pour in.

Spectroscopy

Much of what we know about the composition of matter is deduced from spectroscopy. Spectroscopy is the technique where either (i) electromagnetic radiation is applied to a system and absorption of the radiation as a function of wavelength (energy) is measured, or (ii) other energy is applied to a system and the electromagnetic radiation emitted by the system is measured. This system works because of the requirement for energy to be conserved instantaneously and the fact that atomic and molecular systems are quantized (i.e., restricted to certain specific energy levels):

$$\Delta E = h\nu = hc/\lambda$$

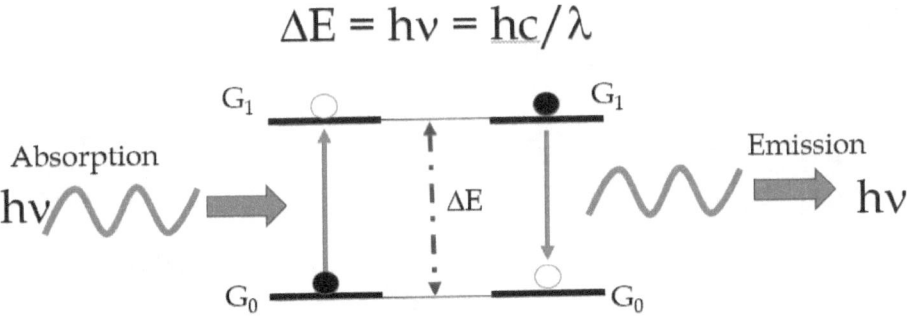

It turns out that certain physical systems tend to fall into various energy ranges.

Magnetic Moments in an External field: *Radio/TV*

Molecular Rotations: *Microwaves*

Molecular Vibrations: *Infrared*

Reversible Electronic Transitions: *Visible*

Breaking Chemical Bonds: *Ultraviolet*

Inner Atomic Electronic Transitions: *X-Rays*

Nuclear Transitions: *Gamma Rays*

For example, here is a typical infrared vibrational spectrum (formaldehyde, $H_2C=O$ in the gas phase):

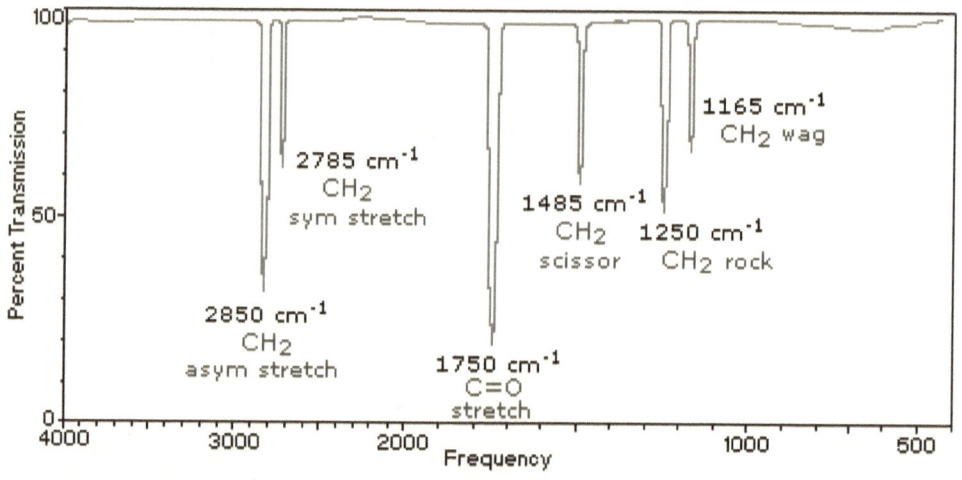

Source:
https://www2.chemistry.msu.edu/faculty/reusch/VirtTxtJml/Spectrpy/Infr aRed/infrared.htm

For a simple molecule like this, it is possible to identify specific stretching and bending modes characteristic of the molecule and assign them to specific absorptions. If the composition of the tested sample is not known, much can be deduced about its structure from these spectra.

Color Vision

One of the amazing things we can do is detect electromagnetic waves, we can even distinguish wavelengths. Well, at least we can do that within a narrow range of wavelengths between 350 and 700 nm. Here is how it works. You are probably familiar with the fact that the eye works a lot like a camera focusing light that enters on to the back section of cells called the retina.

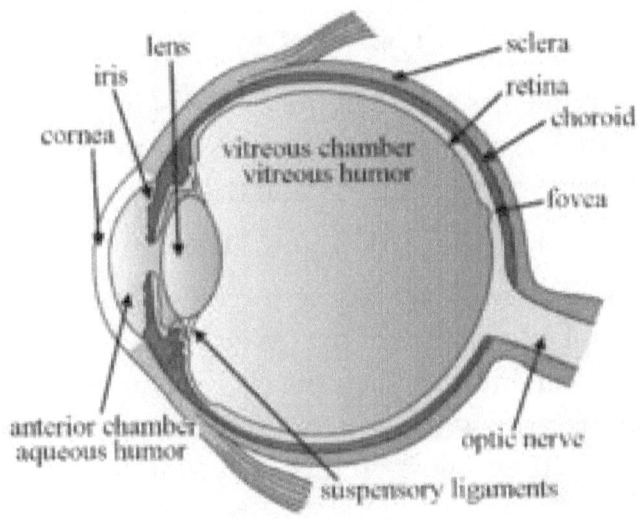

Source: Wikipedia

Many cells on the retina are attached to nerve cells that carry nerve impulses into the brain and create a mental image of what is seen. The key chemistry involves a molecule called **11-cis-retinal** that has a series of double bonds that are tuned to absorb energy from the segment of the electromagnetic spectrum (350 nm-700 nm) where the photons carry enough energy to excite a C-C pi-bond (170 to 340 kJ/mole), but not enough energy to break a C-C sigma-bond (>350 kJ/mole) (see below).

When the photon is absorbed, the pi-bond is broken and the molecule spontaneously rotates from the Cis to the Trans configuration:

$$\textbf{Cis-XYC=CXY + hv} \rightarrow \textbf{[XYC-CXY} \rightarrow \textbf{XYC-CYX]}$$

$$\rightarrow \textbf{Trans-XYC=CYX}$$

While 11-cis-retinal fits nicely in the cavity of a protein molecule called an **opsin**, the trans-form is expelled from the opsin and initiates a chain of nerve transmission that inform the brain of light.

The ability to distinguish wavelengths comes from the fact that there are a number of different opsins and they provide slightly different environments in which the cis-retinal resides. Thus, the opsin is what determines the color. Humans have three opsins and we thus see three colors. If an opsin is mutated, we are color-blind (for that opsin). The trans-retinal is converted back to the cis-form by an enzyme (RPE65) using energy provided by ATP.

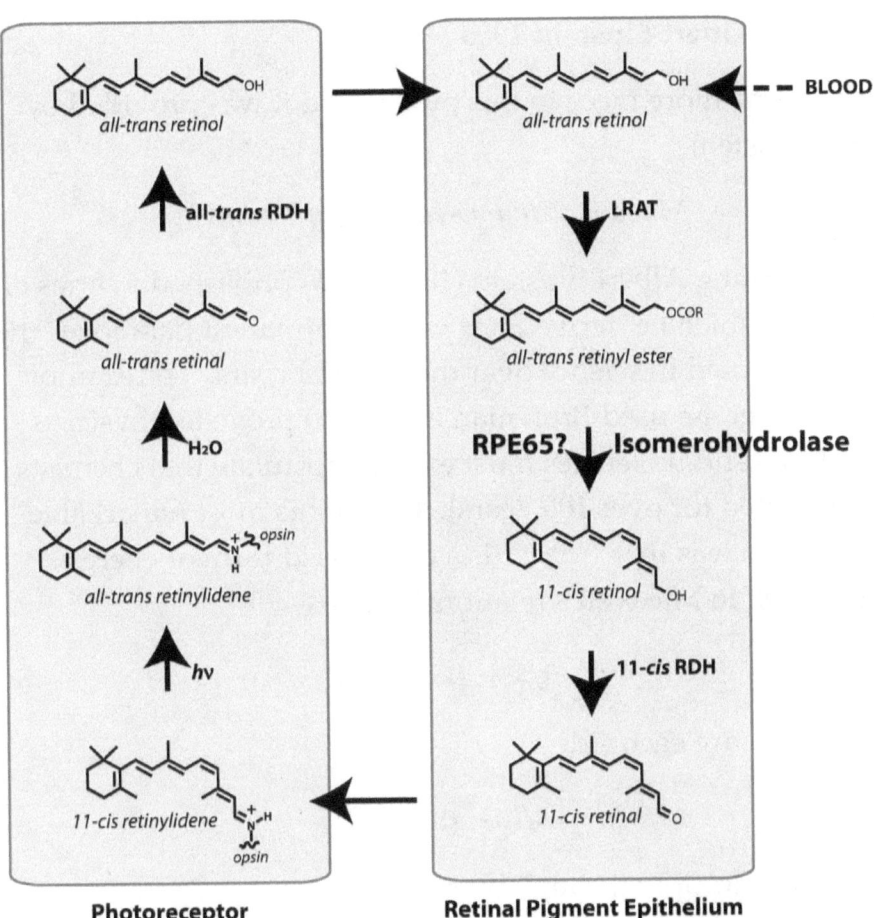

Photoreceptor   **Retinal Pigment Epithelium**

G. Moiseyev, Y. Chen, Y. Takahashi, B. Wu, and J. Ma. 2005.
RPE65 is the isomerohydrolase in the retinoid visual cycle.
PNAS. 102(35):12413-12418.

1.5 Albert Einstein 1905

There is one more piece to this puzzle and it was provided by Albert Einstein.

Mass is Condensed Energy E = mC²

In 1905, young Albert Einstein (1879-1955) published a series of papers in which he derived answers for apparent paradoxes that had been noted in science over the last 100 years. For example, in one paper, he used Brownian motion to prove to physicists that atoms and molecules must exist (a postulate that chemists had accepted for over 100 years). One of his most remarkable conclusions was that "mass" is a condensed form of energy. Going back to Maxwell's relationship:

$$1/(\varepsilon_0 . \mu_0)^{1/2} = C$$

We can square each side:

$$1/(\varepsilon_0 . \mu_0) = C^2$$

And if we multiply through by mass (m)

$$m/(\varepsilon_0 . \mu_0) = mC^2$$

mC^2 has the dimensions of energy,

$$E_{photon} = m/(\varepsilon_0 . \mu_0)$$

Hence the *energy of a photon of electromagnetic radiation*: Electromagnetic radiation must have three components:

energy, a **magnetic vector** and an **electric vector**

And the photons can be in equilibrium with matter, provided that the mass, magnetic and electric properties are conserved.

Matter ⇆ Electromagnetic Radiation

Of course, a lot of energy is packed into a small amount of matter: $E = mC^2$; this was hard for classical physicists to believe and it was not until radio activity was found to involve subtle changes in mass, that the importance of Einstein's equation was appreciated. In order to have enough energy to produce electrons you must have a high energy gamma ray (γ) see figure below.

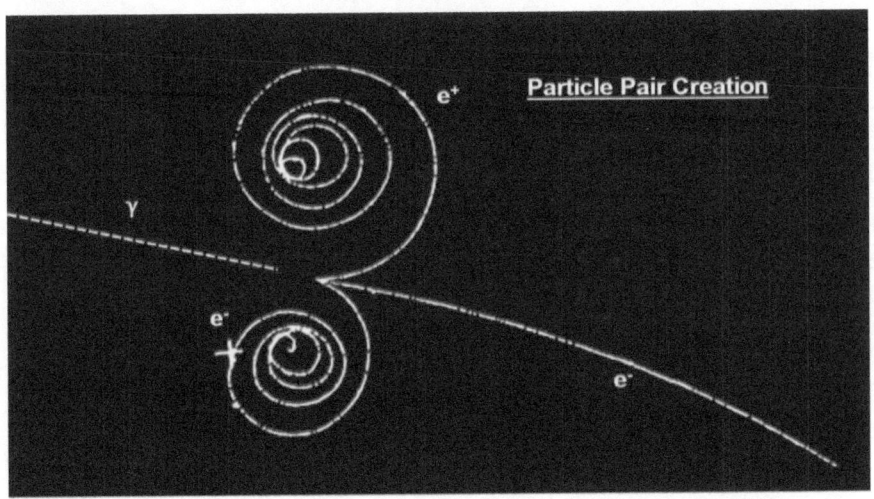

Source:
https://i.ytimg.com/vi/GFxxqgvZVY/maxresdefault.jpg

To conserve the electrical charge both positive and negatively charges particles must be formed and they must have opposite magnetic moments. We thus observe two particles (e+ and e-) spiraling in opposite directions in a magnetic field in a cloud chamber. The gamma ray is coming in from the left, and strikes and electron that carries away kinetic energy to the right while the new particles (with little kinetic energy) spiral near the point they were created.

Photoelectric Effect

Fortunately for Einstein, one of his papers addressed an observed experiment and potentially could be tested. This observation had actually come from the work of Heinrich Hertz (1857–1894), who had observed (1887) that the electrons (cathode rays) emitted by cathodes in a vacuum chamber were facilitated by the slightest amount of short wavelength radiation; but even lengthy exposure to intense long wavelength radiation had no effect. Moreover, while holding the electrodes at a constant potential (voltage difference), the kinetic energy of the ejected electrons increased linearly with the frequency of the light used to stimulate them (see graph).

Hertz had observed this effect and reported it, but had not pursued it. Einstein argued that the effect was completely consistent with Planck's postulate:

$$E_{photon} = h\nu$$

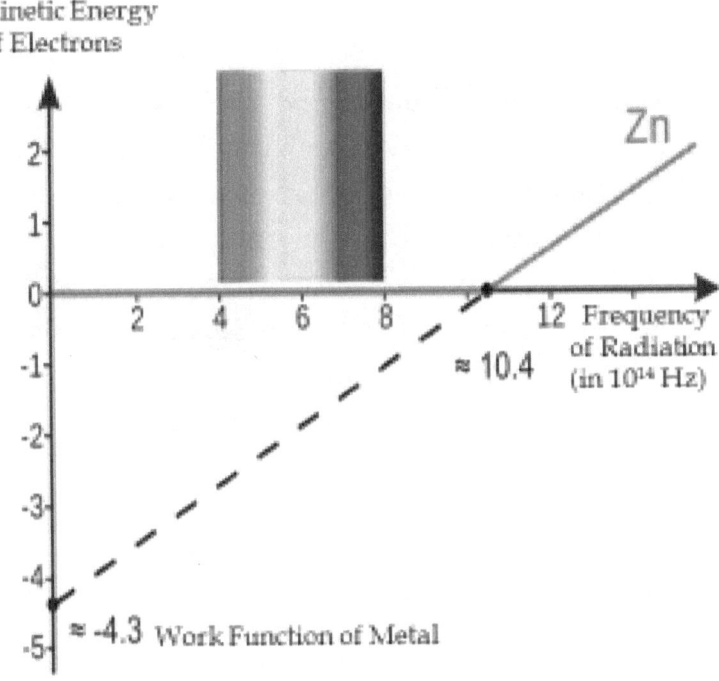

Modified from figure by Klaus-Dieter Keller, sourced from Wikimedia Commons.

And that the slope of the kinetic energy of the ejected electrons was a function of the frequency of the applied light was equal to Planck's constant.

When Planck read the paper, he was, of course, delighted. His criticized idea was confirmed. Einstein had shown a manifestation of Planck's equation that made perfect sense and had experimental roots. But Einstein had not done quantitative experiments.

Interestingly, Robert A. Millikan (1868–1953) was a very capable experimentalists, and he did not think that Einstein or Planck were correct. Thus, Millikan set out to test the theory and after substantial effort, failed to prove it wrong. The result of all of this was that Planck (who was a very well-known scientist) literally came to the patent office where Einstein was working at a menial job and offered him an important academic post in his university. It helps to have friends in high places.

Politics and Prejudice in Science

This promotion brought attention and some envy to some of Einstein's more controversial work. In particular, classical physicists ridiculed Einstein's theory of special relativity. The fact that experimental work by British scientist Arthur Eddington (1915) confirmed one of the most important predictions of the theory was tempered by the fact that Einstein was a Jew and had not supported Germany during the First World War (1914-18). As the issue was coming to a head in 1921, an obscure young man named Wolfgang Pauli (only 21 years old) wrote a readable summary of the theory of relativity. Naturally, Einstein praised the popular book (he hoped the book would help him win the Nobel Prize) and greatly enhanced the recognition of Pauli (who was actually a bully and never (in my opinion) did anything very useful).[18] Nonetheless, the Nobel Committee did not award a prize in physics that year but did award Einstein a prize in 1922 for the photoelectric effect…which was much less controversial.

2.0 Atoms and Ions

2.1 Early Atomic Theory

The ancient Greeks had two hypotheses about what matter is made from. Unfortunately, the idea that "matter is composed of many small particles arranged in different ways" lost out to the idea that matter was composed of air, earth, fire and water. It took more than 2000 years get back on the right track, but by the late 1700s, several scientists, especially Joseph Priestley (1733 – 1804) in England and Antoine-Laurent de Lavoisier (1743–1794)[19] in France were on the right track. Lavoisier is particularly important because he overturned the "air, earth, fire and water" model with careful measurements of mass. Indeed, he

---

[18] Pauli is given credit for postulating the neutrino, which I doubt actually exist. There are alternate explanations for the observations. He won a Nobel Prize for the "exclusion" principle which will be explained below.

[19] Lavoisier was an armature scientist and a professional a tax collector. This career cost him his life on the guillotine in the French Revolution.

introduced to chemists an important principle of chemistry: *In a chemical reaction, exactly the mass that goes into the reaction must come out of the reaction* although the form of the materials may be completely altered.

Electricity

In Italy, Alessandro Giuseppe Volta (1745–1827) invented the "voltaic pile" (literally a battery) that was capable of producing a continuous current in 1800. One of the first discoveries[20] was that when the electrodes from this battery were placed into water, two different gases (oxygen and hydrogen) were produced at the two electrodes:

$$2\ H_2O \rightarrow 2\ H_2 + O_2$$

Electricity was not understood, but apparently it had something to do with bonding of atoms to form compounds.

Early ideas of Chemical Bonding

Humphry Davy (1778–1829) soon learned of the technique and applied it to "potash" and "soda" respectively isolating the active metals potassium and sodium in 1807. The trick to breaking down metal salts and oxides is to melt them so that a current could flow through them. Typically, salts (e.g., common table salt, NaCl) can also be dissolved in water and the solutions

---

[20] William Nicholson and Anthony Carlisle.

conduct electricity. If electricity is passed through a solution of sodium chloride chlorine gas (Cl_2) is produced.[21] It was assumed (incorrectly) that in solution the compound existed as "NaCl" but that the electric current somehow broke the bond.

In the early 1800s, the idea of *atoms* was accepted by chemists, but no one knew anything about their structure. Atoms were obviously very small. The only physical law that chemists were aware of that seemed to apply to this situation was the observations by Coulomb of the attraction of electrically charged particles. Thus, the idea that *atoms were somehow positively charged particles that were glued together by negatively charged electricity* was the dominant view of chemists in the early 1800s.

Isolation and Identification of Elements

From antiquity, a number of metals have been known. Of course, these included gold and silver. Metallic copper was available in some places; lead and mercury rounded out the elements that were likely known in various locations. Iron (in the form of meteorites) was also known in Egypt, because the meteorites did not rust away in desert sands. By the time that Lavoisier got us back on the right track, these metals were fairly

---

[21] Carl Wilhelm Scheele (1742-86) had produced chlorine gas in 1774 from a different reaction, but it was not recognized as an element until isolated by Davy by electrolysis in 1810. Scheele had also isolated oxygen in 1772, but did not publish his results.

obviously either pure elements or perhaps alloys (mixtures) of related metals.

Priestly discovered oxygen (1774) and Lavoisier correctly identified it as an element. Their method of isolation was to heat red mercuric oxide at high temperature to produce a gas (O_2) and the metal (Hg):[22]

$$2\,HgO \rightarrow 2\,Hg + O_2$$

It is relevant that in the late 1700s, heat (a.k.a., phlogiston) and light (i.e., Newton's particles) were considered to be material substances and were included as elements by Lavoisier. As seen above, the nature of light as a wave was soon resolved in the early 1800s.

---

[22] This is a chemical equation. By convention we use the modern symbols that have been adopted to represent the elements (e.g., Hg). Compounds (of two or more elements in chemical combination) are represented by writing the symbols together with subscripts, where appropriate, to represent the relative number of atoms (e.g., H_2O) There are seven elements (H, O, N, F, Cl, Br, I) which are recognized to occur as diatomic molecules in their normal state under ambient conditions. By convention these diatomic molecules are represented in chemical equations as diatomic elements (e.g., O_2). All other elements are represented by the simple symbol (e.g., Hg). In keeping with Lavoisier's teachings, *all the same atoms must appear on both sides of the equation.* And by convention, balanced equations have *lowest-integer multipliers*, where "1" is assumed. The arrow indicated the direction of the reaction: reactants → to products.

Phlogiston theory was more difficult to dislodge; and at one point, hydrogen was thought to be the substance that was responsible for heat. Part of the problem with understanding heat was that in the early 1800s steam power became widely used and was in the hands of engineers who were not really concerned with a modern understanding of matter. They were very happy to treat heat as a fluid that flowed from place to place as their steam engines drove the industrial revolution and made them rich. There are, still today, very distinctly different approaches to thermodynamics by engineers and molecular scientists (e.g., chemists, see Parris, Thermodynamics 3rd Ed., 2019).

The idea was growing that elements were composed of one unique type of "ultimate chemical particle" and that most things in everyday life are compounds of these elements. This idea was a modern version of the ancient Greek idea of atoms. Thus, it followed naturally that elements were composed of only one type of atom.

John Dalton (1766–1844) created a list of the known elements in 1808. Dalton's 1808 list was soon obsolete as Davy and other chemists quickly applied electric current to dissociate salts of many more metals including Na, K, Ca, Sr, and Ba. At this time, no one knew how many elements there were or how they were related to one another.

2.2 Electricity and Ionic Bonding of Atoms

Electrostatic Bonding

If atoms are envisioned as rigid charged spheres, we can easily apply Coulomb's law to calculate the energy that holds them together. We call this energy the *bond energy*.

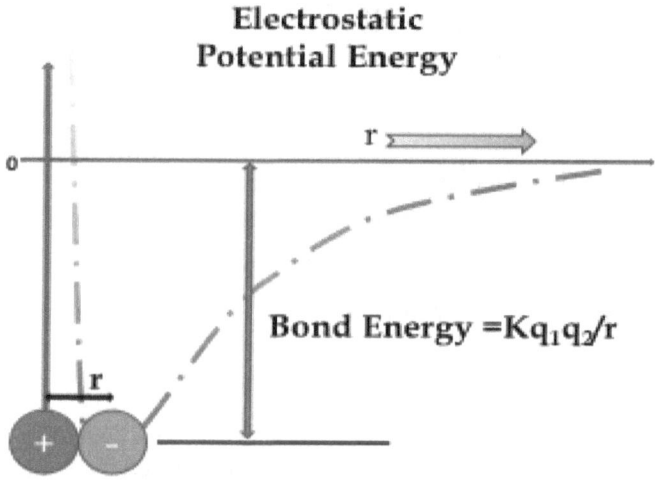

We define infinite separation of the ions ($r = \infty$) as the situation where there is no potential energy in the system. Notice that as the ions with *opposite charges* approach one another (i.e., as **r** becomes smaller), the potential energy is negative. When the rigid ions come into contact, they begin to repel and the potential energy increases abruptly. In the case of sodium chloride (NaCl), the sodium atom transfers an electron to the chlorine atom to produce oppositely charged ions. The ions approach one another and as they do, the potential energy (dashed line) of the system decreases until the ions touch.

Ionic Radii
Na⁺ 1.16 Å   Cl⁻ 1.67Å

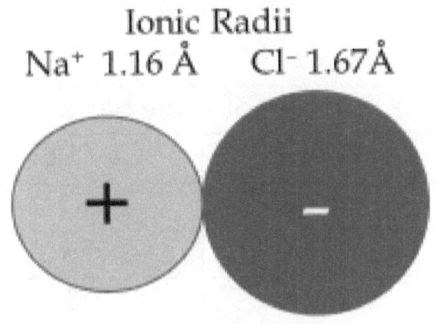

$$E = (1.39 \times 10^3 \text{ kJ Å/mole})(+1)(-1)/2.83 \text{ Å}$$

$$E = -491 \text{ kJ/mole}$$

The figure above shows how Coulomb's law can be applied quantitatively to calculate the ionic bond energy. In this case, a sodium ion (Na⁺) and a chloride ion (Cl⁻) are assumed to be in contact forming an ion pair. The distance between the center of the charges (r) is the sum of the ionic radii.

If we consider an entire crystal (involving billions of ions), the energy of the entire system (e.g., the "lattice energy" holding an ionic crystal together) can be calculated from application of Coulomb's law to every pair-wise interaction. Notice that the interactions of like-charged particles introduce positive contributions to the calculation.

For sodium chloride the experimental value of the lattice energy is 786 kJ/mole. (By convention, *lattice energy* is defined as the energy needed to be put into the system to break the bonds and thus is positive.)

Lattice Energies of some Salts (kJ/mole)			
Ion/radius (Å)	F-/1.19	Cl-/1.67	Br-/2.06
Li+/0.90	1036	853	807
Na+/1.16	923	786	747
K+/1.52	821	715	682

The Electrostatic Model of Molecules

A theory was evolving in the early 1800s that all atoms had an inherent charge (either + or -) and this was assumed to account for their mutual attraction (bonding). For example, metals (Na, K, Mg, Ca) formed compounds with non-metals (O, Cl, Br). Since water was electrolyzed to produce hydrogen and oxygen, hydrogen must behave like the metals with a positive charge in compounds. The non-metals must have negative charges in compounds. The main contributor to this theory was Hermann Kolbe (1818-1884) in Germany (circa 1832).

The idea soon arose was that certain combinations of atoms had unusual stability (through electrostatic attraction) and these were called "radicals" (a term borrowed form botany for "root" meaning the fundamental part).[23] Recall that electrostatics do not have any preferred orientation. *Thus, with no forces other than electrostatics to hold atoms and ions together, the only model of radicals that had actual theoretical support was the idea that radicals were nothing more than unusually stable clumps of ions.* Kolb viewed organic radicals rather like a pile of marbles held together by glue or perhaps a bunch of grapes all bound to a central stem.

Photo Agne27 attributed to Don Kasak, Source: Wikimedia Commons

While the basic idea was correct, Kolb (who became a very well-known chemist late in his career) stood in the way of any thought that these clumps of atoms could have any preferred three-dimensional shape. This idea was not proven wrong until crystallography, and a theory to explain it was introduced in

---

[23] Today, chemists use the term "radical" to identify complicated parts of organic molecules (which are typically not involved in the reaction of interest).

1932. It also stood in the way of the idea that salts (i.e., ionic compounds) can exist as independent ions in solution.

What Kolb did not understand is that water (the most common solvent for salts) is made up of very *polar molecules* of H_2O.

Ions in Solution

It was not until 1887 that Svante Arrhenius (1859–1927) recognized that in solution most salts and acids at least partially dissociate into *solvated ions*:

Crystal ⇆ Ion Pair[24] ⇆ Solvated Ions

This is an "equilibrium," which chemists represent with arrows going both ways (⇆). Equilibria involving ions and ion clusters in solution are generally very fast (they are established almost instantaneously) as the salt or acid dissolves in water.[25]

You can represent a polar molecule as a dipole moment (see above). Although these dipole-ion interactions are not as strong as ion-ion interactions, an ion can be surrounded by many

---

[24] Other clusters of ions are common. For example, ion triples can form: **(+)(-)(+)** The composition of solutions depends upon the sizes and charges of the ions, the concentrations of the ions and the polarity of the solvent.

[25] Chemical systems tend to the most **stable** arrangements of atoms, molecules and ions. If this is reversible and very fast, chemist say the system is **labile**. The opposite of stable is **unstable**; the opposite if labile is **inert**.

dipoles. This arrangement of dipoles and ions produces a more stable arrangement than an ion being surrounded by a small number of ions.

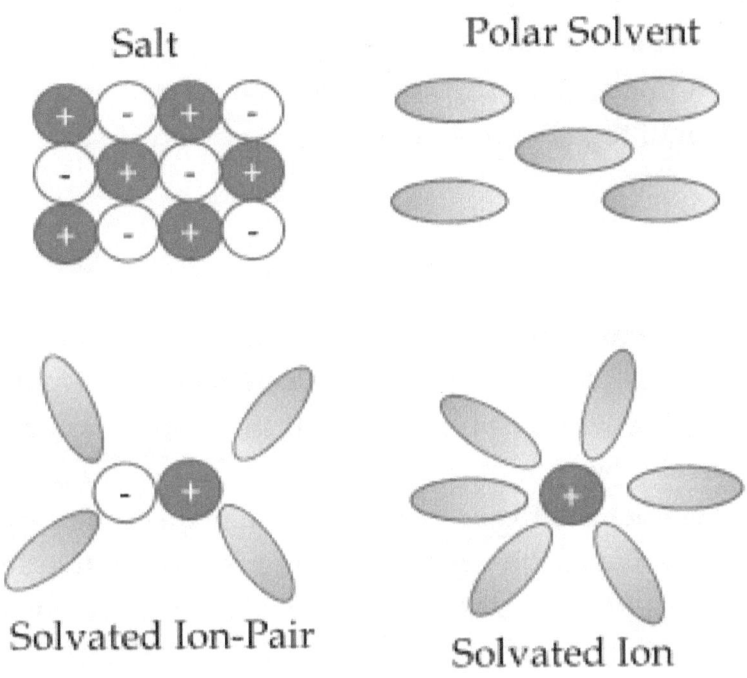

We now know that even in the solid state, salts are typically composed of independent **ions**. For example, some salts are formed by atomic ions (Na^+ and Cl^-) while some ions are charged molecules (NO_3^-). In the presence of a polar solvent, the salt can dissolve to produce a solution containing ions and ion pairs.

2.3 Pieces of the Atomic Puzzle

During the 1800s and early 1900s some interesting observations were made that did not immediately lead to a better understanding of the structure of the atom. But these pieces were brought together in an unexpected way in 1912.

The Periodic Table

Dmitri Mendeleev (1834-1907) was one of several people who developed schemes for organizing the elements. His motivation was simple; he was writing a textbook and wanted some way to systematically present the information to students. Thus, he grouped elements into families with similar chemical properties (e.g., Li, Na, K) and then ordered the groups by relative atomic mass. At this time, elements were still being discovered and no one knew how many elements there might be. His table was presented in 1869; and because he used it to predict new elements and accurately predicted their properties, he is generally given credit for the table.

The order of the elements in the original table 1, 2, 3, etc. was based on the apparent atomic masses, with numbered spaces left where it appeared that new elements would be discovered. At the time, these numbers had no physical significance. Today, we recognize several facts that were not apparent until the mid-

1900s. In particular, the order numbers are now recognized as "atomic numbers" characteristic of the elements and we now know that the atomic masses are actually weighted averages of the *natural abundance* of various atomic *isotopes*. Isotopes of an element all have the same chemical properties (atomic number), but they have different atomic masses. We will explain why this happens below.

Atomic Spectra

Equipped with high quality glass prisms, John Herschel and others came to the realization that the *line spectra* produced when an element was placed in a flame were characteristic of the element. And, in 1859, Kirchhoff was able to state definitively that elements absorbed and emitted light with the same pattern of wavelengths. These studies were aided substantially by the invention of a light source that used gases generally free of metallic elements, by Kirchhoff's colleague Robert Bunsen (1811–1899). By 1860, Bunsen and Kirchhoff had identified two new alkali metals by their line spectra.

Atomic Spectrum of Potassium

Author: McZusatz; Source: Wikipedia

In Sweden Johannes Rydberg (1854–1919) was a physicist who became interested in the periodic table in which Mendeleev had recently organized the elements. He felt there must be mathematical relationships among physical properties that would shed light on the structure of atoms. His interests in the 1880s moved to the atomic spectra, but he was working with the spectra of many metals and was not able to deduce any fundamental relationship although there seemed to be several series of lines. Information about his hypothesis began to circulate in Europe; and in Switzerland, a young physicist approached a senior mathematician Johann Jakob Balmer (1825–1898) for help on the problem. But they did not have access to many spectra and Balmer worked only with hydrogen. For reasons that will be clear below, this was the perfect element to try (because it only has one electron).

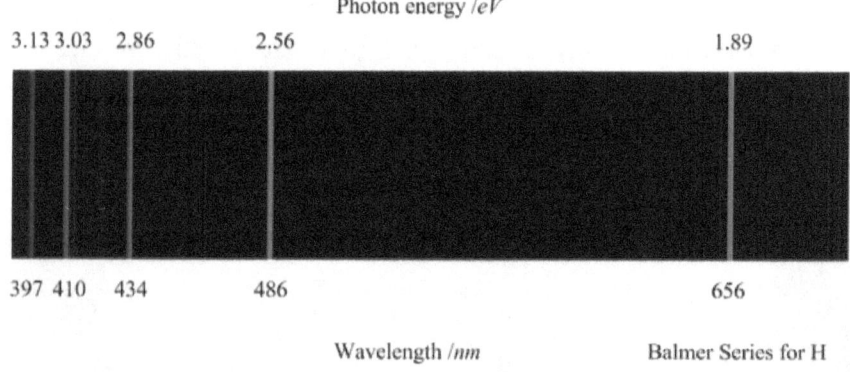

Author: Franky; Source:
https://franklyandjournal.wordpress.com/2016/07/18/hydrogen-spectrum/

The hydrogen spectrum was very simple and Balmer quickly developed a mathematical equation that accounted for the lines in the visible spectrum and predicted lines that had not previously been observed at longer and shorter wavelengths.

The ability to predict new lines gave great credibility to Rydberg's hypothesis and Balmer's results were published in 1885.

$$\lambda = h \, \frac{m^2}{m^2 - n^2}$$

His relationship (above) had two parameter (m and n) that were incremented as integers (1, 2, 3, 4…). In the case of the visible spectrum of hydrogen, m was 2 such that $m^2 = 4$.

As soon as Rydberg saw what Balmer had discovered, he was able to reformulate his ideas and developed a series of more complicated equations in terms of $1/\lambda$, which we now know (thanks to Planck) is proportional to photon energy. In the case of hydrogen, Rydberg's equation reduces to the form shown below:

$$\frac{1}{\lambda} = \frac{4}{h} \left(\frac{1}{n_1^2} - \frac{1}{n_2^2} \right) = R_H \left(\frac{1}{n_1^2} - \frac{1}{n_2^2} \right)$$

In 1896, Pieter Zeeman (1865–1943) observed unexpected splitting of atomic spectral lines when the discharges originated in strong magnetic fields. But the basis for this splitting could not be understood until much more information about atomic structure was known.

The Structural Hypothesis

Recall that Hermann Kolbe relied exclusively on the electrostatic properties of atoms to explain compounds. And, he was correct that was the only physical force that was *known* to hold atoms together. Nonetheless, Louis Pasteur (1848) and August Kekulé (1865) and others began to think in terms of atoms being arranged in three-dimensional space in certain specific geometries. In particular, Jacobus "Henry" van 't Hoff, Jr. (1852-1911) published a pamphlet in 1874 that explained the phenomenon of optical activity based on tetrahedral carbon (which will be explained below). In addition, organic chemists were starting to draw structures to explain chemical reactions that required "radicals" to have structures that were not explained by pure electrostatics.

2.4 Advances in Atomic Theory

The Electron and the Atomic Model of J.J. Thomson

In the late 1800s, it was hard to rationalize chemical reactions and the existence of *molecules* (i.e., clusters of atoms that were not dissociated into individual ions by polar solvents). As mentioned above, by 1887, Arrhenius had convinced people that salts existed in water as separate ions. Why were some clusters of atoms resistant to solvation and others not? Obviously, there was interest in the actual structures of atoms.

It had been discovered (in the early 1800s) that if a very high voltage is applied to electrodes, current could be caused to flow through the air as a spark. If the electrodes were placed in a glass tube so that the composition of the gas and the pressure of the gas could be changed, it was found that under some conditions, the entire tube could be caused to glow in exotic colors. The ability of scientists to reduce the pressure inside the glass tube gradually improved with invention of better and better vacuum pumps. In 1875, William Crookes (1832–1919) managed to evacuate a tube to the extent that current flowed steadily in a single defined line from a cathode to an anode. This was called the "cathode ray" tube.

J.J. Thomson (1856-1840) was a physicist who recognized the potential of this tube to study the properties of electricity. In 1897, Thomson studied the deflection of the cathode ray by magnetic fields and concluded (based on electromagnetic principles) that the ray was actually a stream of very small negatively-charged particles, which were soon called *electrons*. Electrons were homogeneous in mass and charge; and it did not matter what metal was used for the electrodes, they all produced the same particles.

Using basic physics, Thomson was able to directly calculate the ratio of *mass over charge* (m/e); but he was not able to produce accurate masses of the electrons. Nonetheless, it was obvious that the electrons were a part of the atom and accounted for only a small fraction of the mass. Thus (still relying solely on electrostatic concepts of Coulomb), Thomson proposed a model of the atom in which the electrons were embedded in a massive positive charge (1904).

X-Rays

In 1895, Wilhelm Röntgen (1845 – 1923) observed that when high-kinetic-energy electrons collided with the anode of a cathode ray tube, a type of electromagnetic radiation that had shorter wavelengths that ultraviolet light was produced. Because these rays were of such high energy, there is no mechanism for them to be absorbed by atoms of the first or second row of the periodic chart (e.g., C, H, N, O) which compose most of the soft tissues of the human body.

These waves are however impinged by heavier elements (Ca, Sr, P, S and trace transition metals) found in bones. The medical implications were immediately realized and within weeks of this discovery doctors were making x-rays of patients to diagnose broken bones and related diseases.

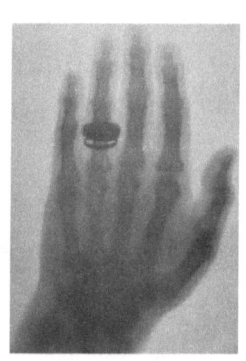

Röntgen's original x-ray →

Radioactivity

Incidental to his study of phosphorescence of uranium minerals, Antoine Henri Becquerel (1852 – 1908) discovered that some sort of radiation from the mineral penetrated protective paper wrappings and left images on photographic plates. He published his observations merely a day ahead of Silvanus Phillips Thompson (1851 – 1916) and has received credit for the discovery. This happened to be shortly after x-rays were discovered by a similar technique and the initial assumption was that this was another form of x-rays. But subsequent work (published by 1899) proved that at least some of the radiation from uranium salts was similar to cathode rays (electrons) but the electrons had more penetrating power.

This was a new field and it attracted Marie Skłodowska (1867 – 1934) who had come from Poland to study in France. Marie had married Pierre Curie (1895) and he happened to have developed a device for measuring weak electrical currents in air caused by the ionization as electrons passed through the air. Thus, the Curies had a convenient way to quantitively measure the rate at which ionizing radiation was being emitted from a sample. Sparked by Marie's desire to isolate the source of Becquerel's high-kinetic-energy electrons, they began extracting fractions form the uranium minerals and soon discovered two things: (i) There was very little of whatever was releasing the electrons in the sample and (ii) it was many orders or magnitude more reactive than pure uranium metal. Eventually, Marie and Pierre isolated Polonium and Radium in samples large enough to be studied.

About this same time, Ernest Rutherford (1871 – 1937) a student of J.J. Thomson, also became interested in radioactivity.

Rutherford carefully examined the radiation from uranium and thorium and found it could be classified as *alpha* particles which had very little penetrating power (they are effectively only released from the surface of a material) and *beta* particles (which were like high energy cathode rays, electrons).

Rutherford also discovered that some radioactive derivatives isolated from known radioactive elements decayed very rapidly producing still other radioactive materials. From these observations, he invented the concept of a half-life ($t_{1/2}$, *the time that it takes for half of a radioactive material to disappear*).

Frederick Soddy (1877 – 1956) joined Rutherford at McGill university in Canada in 1900. During their brief association, they realized that these transient radioactive materials were actually unstable versions of several different elements….one element was being converted into another element (*transmutation!*).

With the isolation of pure radium, it became obvious that heat was being released in radioactivity. But it was not until the 1940s that it was understood mass as being converted into energy following the prediction of Einstein's equation $E = mC^2$.

In addition to alpha and beta radiation, Rutherford eventually discovered that very-high-energy electromagnetic radiation (*gamma* (γ) radiation) is often released in nuclear reactions. The reason for this is conservation of energy. When one mass is converted to other masses, any excess energy that is not otherwise accounted for is emitted in the form of a gamma-ray of electromagnetic energy. As noted above, electromagnetic radiation is nature's way of *making change* (to ensure conservation of energy).

These observations were important to the development of the understanding of the nature of the atom, but because they involve the nucleus, they have very little direct relevance to *chemistry*. Thus, we will spend very little time on the topic of radioactivity and nuclear energy.

Isotopes

We have already mentioned that atoms of the same element can have slightly different masses. This was first suspected by Soddy who realized that certain radioactive elements went through a sequence of steps (e.g., an alpha particle[26] was lost; followed by loss of two beta particles). Clearly, there must have been a change in mass, but the chemical behavior (i.e., elemental identification) was unchanged. The basis for this behavior was not understood until about 1930.

You must have enormous respect for the scientific honest of Rutherford and Soddy who were reporting that they started with an element, which changed into a different element and in a few days returned into the original element. Their devotion to

---

[26] After moving to Manchester, England, Rutherford and Thomas Royds demonstrated that alpha particles are actually helium nuclei (1907).

careful experimentation and honest reporting outweighed their "common sense."[27]

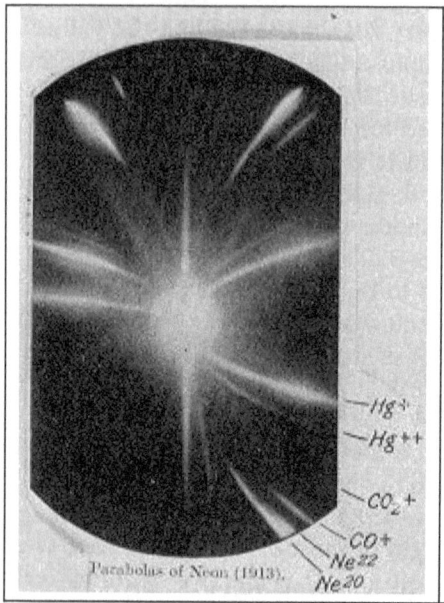

Parabolas of Neon (1913).

However, the existence of isotopes was proven during work by J.J. Thompson (1913). Thompson developed techniques for accelerating positively charged ions and launching them into magnetic fields where they were deflected into an arc, which depended on their mass. In experiments with neon, Thomson discovered that it had two isotopes with masses 20 and 22.

Supernova of massive stars likely produce a wide range of isotopes of all elements. However, those with short half-lives decay away over time leaving what we call the *natural abundance* of (relatively) stable isotopes. Potassium, thorium and uranium have long-lived radioactive isotopes and survive to this day. In addition, the decay of these isotopes maintains a trickle of naturally occurring isotopes with short half-lives (like radium and radon).

---

[27] I do not grade lab students on getting the "right answer or a high yield." I grade them on reporting what they observe. Technique can be taught, once integrity is gone it is hard to get back.

2.5 The Nuclear Atom

The Geiger-Marsden Experiment

By 1909 Rutherford was a well-known physicist and had a large research group. As a student of Thomson, he supported the "plum pudding" model of the atom that Thomson had proposed. While Rutherford was preparing to receive the Nobel Prize, he assigned his post-doc Hans Geiger (1882 – 1945) and a graduate student Ernest Marsden (1889 – 1970) to perform experiments to see how alpha particles passed through a thin gold foil.[28]   The actual sequence of events and arrangement of the apparatus is usually greatly simplified in textbooks (for a true account of this research I recommend my book *The History of Atomic Theory*, 2018, paperback 2019). Suffice it to say that through a series of experiments Geiger demonstrated that alpha particles (known to be heavy positively charges particles) were reflected from the interior of the gold foil (not just from the surface) as well as other metals and that the repulsion could not reasonably be magnetic.

Geiger stated the most interesting conclusion from this work[29] was as follows:

---

[28] Geiger had already done some experiments (1908) with alpha particles reflected off of a metallic surface.

[29] Geiger, H. 1910. The scattering of the alpha-particles by matter. *Proceedings of the Royal Society. A* 83:492-504.

*"It is also of interest to refer here to experiments made by E. Marsden and myself (see 'Roy. Soc. Proc.,' A, vol. 82, p. 495, 1909) on the diffuse reflection of the alpha-particles. It was found that some of the alpha-particles falling upon a metal plate appear to be reflected, i.e. they are scattered to such an extent that they emerge again on the side of incidence. It was shown that from gold 1 in about 8000 of the incident a-particles suffers reflection, and that reflection takes place within a relatively thin surface layer equivalent to about 5 mm. of air. According to the curve (fig. 4), the probable angle through which the alpha-particles are turned in passing through this equivalent thickness of gold is only about 1°, and a simple calculation, assuming ordinary probability law, shows that the probability of an a-particle scattered through an angle exceeding 90° is extremely small, and of a different order from that which the reflection experiment suggests. **It does not appear profitable at present to discuss the assumption which might be made to account for this difference.**"* [30]

Rutherford spend a year reviewing the data and doing mathematical modeling before he was prepared to publish his interpretation of these observations:

---

[30] I think Geiger knew what it meant, but Rutherford would not let him say it. Rutherford's motives are unclear. He was a very serious experimentalist and was not inclined to theorize without extensive proof. These are the traits that allowed him to do the tedious work for which he was known. We will see again Rutherford's reluctance to theorize when he meets Bohr and Moseley.

> "*...The scattering of the electrified particles is considered for a type of atom which consists of a central electric charge concentrated at a point and surrounded by a uniform spherical distribution of opposite electricity equal in amount.* With this atomic arrangement, an a or β particle, when it passes close to the centre of the atom, suffers a large deflexion, although the probability of such large deflexions is small.*"[31]

A do not blame Rutherford for carefully considering the implications of this model of the atom before publishing this conclusion[32], because it flies into the face of the idea of electrostatic attraction (Coulomb's law). How could the negative charge and positive charge be separated?

A famous Rutherford Quote: "*If your experiment needs statistics, you ought to have done a better experiment.*"

―――――――――――――――――

[31]E. Rutherford, The Scattering of the a and β Rays and the Structure of the Atom *Proceedings of the Manchester Literary and Philosophical Society*, IV, 55, pp. 18-20. Abstract of a paper read before the Society on March 7, 1911

[32] But I wish Geiger's name had been included as a coauthor. I'm confident that Geiger guessed what was going on (his assumption).

The Rutherford-Bohr Model of the Atom

It happened that while this work was going on a young scientist from Copenhagen, Denmark was a visiting scientist in Rutherford's lab. Niels Bohr (1885 – 1962) had been interested in atomic structure and was well-schooled in quantum mechanics by Max Planck. Bohr began trying to rationalize Rutherford's model of the atom by assuming that the electron (which was known to be much less massive than the nucleus) orbited around the nucleus. There is a connection between the orbital radius and the speed of the particle in classical mechanics. But classical mechanics does not preclude any radius or speed (orbital frequency). It did not take Bohr long to realize that if he assumed that the potential energy of the electron was subject to quantization, a series of distinct orbital energy levels could be rationalized. And working with data for hydrogen (one proton and one electron) Bohr soon showed that he could explain quantitatively the mysterious lines in the atomic spectra (i.e., the work of Rydberg and Balmer) using the Planck's postulate and his derived energy levels. [33]

---

[33] Niels Bohr, N. July 1913. On the Constitution of Atoms and Molecules I. *Philosophical Magazine* Series 6, 26:1-25.
http://www.chemteam.info/Chem-History/Bohr/Bohr-1913a.html

Bohr, N. September 1913. On the Constitution of Atoms and Molecules, Part II Systems Containing Only a Single Nucleus. *Philosophical Magazine.* 26 (153): 476–502.

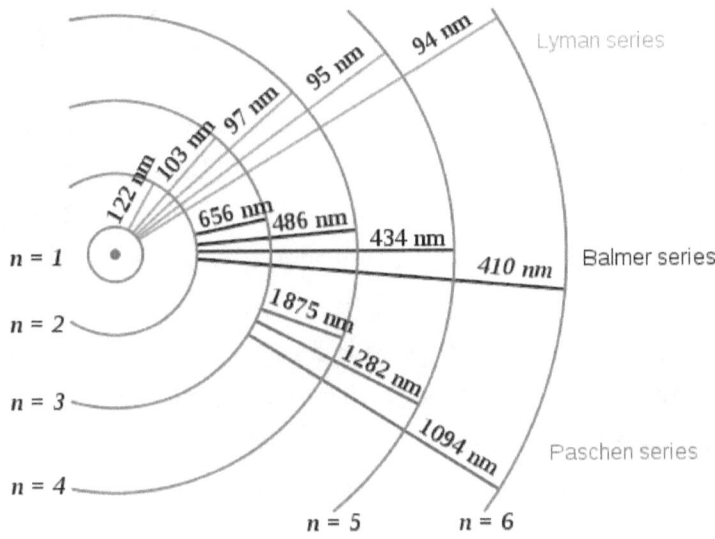

Author: A_hidrogen_szinkepei.jpg (Szdori) derivative work: OrangeDog

Source: Wikimedia Commons

Henry Moseley (1887 - 1915), another student of Rutherford, was collecting x-ray data on metals (late October 1913) and corresponded with Bohr as he prepared his paper (December 1913). Moseley and Rutherford reasoned that x-ray spectra must (in some way) support the ideas of the Bohr model.

Atomic Numbers

As you may recall, x-rays were generated by firing high energy electrons at a metal target. The electron kicks out a core electron and less stable electrons then fall from higher energy states to the vacant positions nearer the nucleus and a high-energy photon is emitted. In 1912, Moseley (with the mathematical aid of Charles G. Darwin) successfully invented a method for quickly capturing the x-ray line spectrum of an element in a photograph, similar to the methods used for UV and visible light. He, thus, had a tool for rapidly determining the x-ray spectrum of a large numbers of elements.[34]   In 1913, He applied this technique to a number of metals and discovered that there is a regular step in the position of the spectral lines, which is evident by plotting the order of the elements in the periodic table (on the ordinate) and the square-root of the frequency (on the abscissa). Not only do these spectra provide a powerful analytical technique for identifying elements (similar to the approach of Bunsen and Kirchhoff); they provided direct information about the nucleus and its charge[35] because the nuclear charge (Z) determines the relative energies of the lowest Bohr orbits:

> *"Now if either the elements were not characterized by these integers* [atomic numbers], *or any mistake had been made in the order chosen or in the number of places left for unknown elements, these regularities would at once disappear. We can*

---

[34] Moseley, H. G. J. 1913. The high-frequency spectra of the elements. *Phil. Mag.* p.1024.

[35] Moseley, H. G. J. 1914. The high-frequency spectra of the elements. *Phil. Mag.* p.703. Mosley was a patriotic young man who joined the British army and was killed at Gallipoli in 1915.

therefore conclude from the evidence of the X-ray spectra alone, without using any theory of atomic structure, that these integers are really characteristic of the elements. Further, as it is improbable that two different stable elements should have the same integer, three, and only three, more elements are likely to exist between Al and Au. ..."

*"Now Rutherford has proved that the most important constituent of an atom is its central positively charge nucleus, and van den Broek has put forward the view that the charge carried by this nucleus is in all cases an integral multiple of the charge on the hydrogen nucleus. **There is every reason to suppose that the integer which controls the X-ray spectrum is the same as the number of electrical units in the nucleus, and these experiments therefore give the strongest possible support to the hypothesis of van den Broek.**[36] Soddy has pointed out that the chemical properties of the radio-elements are strong evidence that this hypothesis is true for the elements from thallium to uranium, so that its general validity would now seem to be established."*

Moseley's discovery cleared the air on a number of topics: (i) It equated atomic number with the identity of an element and identified atomic number as the nuclear charge; (ii) It confirmed both Rutherford's nuclear atom and Bohr's theory of electron configuration; (iii) It confirmed the isotope arguments of Soddy

---

[36] van den Broek, A. 1913. Die Radioelemente, das periodische System und die Konstitution der Atome. *Phys. Z.* 14:32.

and Fajans; and (iv) It opened the question about neutral mass in the nucleus.

The Components of the Nucleus

The nuclei of atoms play little role in the chemistry of elements. But before focusing entirely on chemistry and the electrons, we should take a moment and summarize the importance of the nucleus from a chemist's point of view. Further work by Rutherford's group in Britain, Otto Han and Lisa Meitner in Germany and Marie Curie's daughter in France led to discovery of two particles in the nucleus: protons and neutrons. They have similar mass and chemists rarely care, but physicists would take me to task for the table provided below:

Particle	Mass (AMU)	Charge (e)	Magnetic moment
Electron	0	-1	Large
Proton	1	+1	Small
Neutron	1	0	Minimal

The protons and neutrons make up the *nucleus*, which is very small ($\sim 10^{-15}$ m). Thus, the nucleus has a positive charge equal to the atomic number. The number of neutrons increases the mass of the nucleus and affects the stability of the nucleus, but plays little role in the chemical behavior of atoms. Presumably, when nuclei were produced in supernova, there was a large variety of isotopes (i.e., combinations of protons and neutrons) formed in a random fashion. But what we observe today are the isotopes that are stable or at least undergo nuclear reactions very slowly (with half-lives on the order of billions of years). We would call this the *natural abundance* of isotopes of the elements.

For example, there are three isotopes of hydrogen known:

	Protons	Neutrons	Electrons	
^1H	1	0	1	Stable (99%)
^2H	1	1	1	Stable (1%)
^3H	1	2	1	$t_{1/2} \sim 12$ years

Two of the isotopes are stable but the third (i.e., tritium) only exist as a manmade isotope because if it were formed long ago, it would have vanished over billions of years. These two isotopes (relative abundance is approximately 100 to 1) account for the average atomic mass of hydrogen:

$100 \times 1 = 100$

$\underline{1 \times 2 = \qquad 2}$

$\qquad\qquad$ 102 amu

Thus, 101 particles have a combined mass of 102 amu; and the average mass is $102/101 \approx 1.009$. Actual measured *average* atomic mass of hydrogen is 1.008 amu.

The average atomic masses of the lighter elements are often close to 2 x the atomic number. But for heavier elements the number of neutrons far exceeds the number of protons.

Nuclear decay typically involves loss of an *alpha particle* (two protons and two neutrons) or decay of a neutron into a proton and an electron moving a high velocity (*beta particle*).

De Broglie Electrons as Waves

In 1909 Einstein had published "On the current state of the radiation problem" in which he was trying to rationalize how a wave that is dispersed could deliver a precise complete packet of energy (photon) to a point as in the photoelectric effect. Einstein argued that "corpuscles" could have dual properties of particles and waves.[37] This idea was picked up initially (1924) by Louis de Broglie (1892–1987) to suggest that electrons could be represented by an electromagnetic wave. This wave would need to be in phase to not cancel out, and thus, only orbits with even numbers of wavelengths would be permitted. This was offered as an explanation for the discrete orbits proposed by Bohr. Both Bohr and De Broglie seem to be proposing a model that is essentially similar to the solar system. The problem is the solar system is a disk "flat."

---

[37] Einstein, A. 1909. Zum gegenwärtigen Stand des Strahlungsproblems, *Phys. Zeitschr.* 10:185-193.

2.6 Quantum Mechanics and the Atom

Heisenberg and Schrodinger

It is obvious that atoms are three-dimensional and Werner Karl Heisenberg (1901–1976)[38] soon developed a mathematical representation of three-dimensional waves (called matrix mechanics) that would account for the disposition of electrons in atoms. But this was a very abstract format that was hard to visualize. Fortunately for chemists and all people who are not advanced mathematicians, Erwin Schrödinger (1887–1961) soon (1926) produced a much more tangible representation of the electron distribution around atoms. He applied an energy operator ($\hat{H}$) to a wave function (Ψ) to calculate the energy (E) and the wave function:

$$\hat{H}\Psi = E\Psi$$

This equation appears in most modern chemistry books and few PhD chemists have a clue what it means (so do not feel lost).[39] The results of these calculations are an energy and a wave

---

[38] Heisenberg lead the German attempt to harness nuclear fission during WWII. He won the 1932 Nobel prize for "the creation of quantum mechanics."

[39] I have written a book (Application of Schrodinger's Equation to Simple Physical Systems) currently only available in digital format on Amazon/Kindle, which at least suggests how the method is applied.

function, which represents a wave in three dimensions. That wave function is a mathematical function of four quantum numbers (n, ℓ, $m_ℓ$, m_s):

$$\Psi(n, l, ml, ms)$$

The wave function itself can take on positive or negative values, but the square of the wave function (Ψ^2) is proportional to the *electron density* in space around the nucleus of the isolated atom.

The squared-wave functions are called *orbitals* and they are typically represented in three-dimensions as *a surface that contains a high percentage (e.g., 80%) of the probability of finding the electron in the enclosed space.*

The quantum numbers can be considered as follows:

"n" is the Rydberg-Bohr quantum number which is the primarily determinant of the energy of the electron in the orbital. This quantum number is an integer 1, 2, 3, 4, etc. Because the energy is closely related to the electrostatic potential, it also defines the radius of the orbital, and thus, the size of the orbital.

The second quantum number ($ℓ$) is represented by a lower-case script "L." It is called the azimuthal quantum number and it defines the shape of the orbital. This quantum number can vary from 0, 1, 2, 3...n-1. Such that when n = 1, $ℓ$ can only be 0; but when n = 2 then $ℓ$ = 0 or $ℓ$ = 1.

For $\ell = 0$, the shape is spherical; for $\ell = 1$ the shape has two lobes; and for $\ell = 2$ the shape has four lobes (or the equivalent) as shown in the following figure.

As shown in the figure, when the shape is spherical, there is only one orientation in space; when the shape is essentially linear the orientation can be along the x, y or z axis; and when the shapes are more complex ($\ell = 2$) there can be five orientations.

The orientation is defined by what is called the magnetic quantum number (m_ℓ) [m sub-ℓ]. This quantum number can vary from - ℓ...0...+ℓ. In other words when $\ell = 0$, m sub-ℓ must be zero. When $\ell = 1$, m sub-ℓ can be -1, 0 or +1. When $\ell = 2$, m sub-ℓ can be -2, -1, 0, +1 or +2.

For reasons discussed in section 2.4, you can only place one or two electrons in an orbital.

For historical reasons, the shapes of the orbitals are designated by letters (s, p and d) for the quantum numbers $\ell = 0, 1, 2$, respectively.

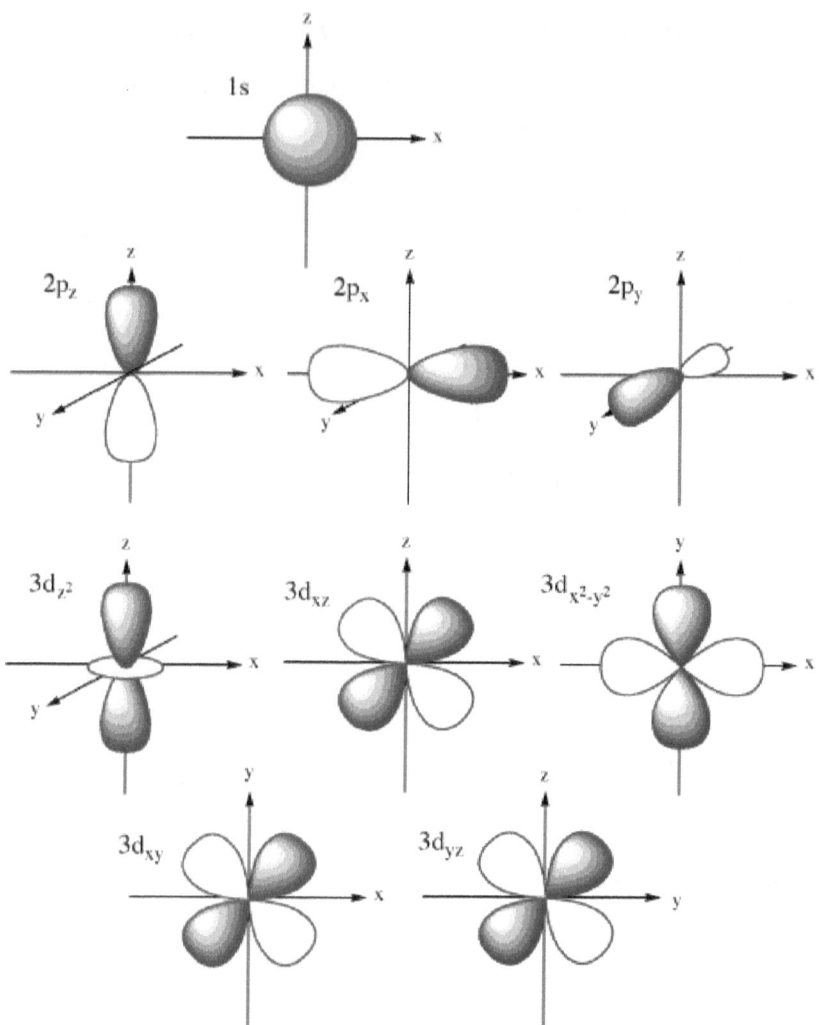

Source: https://www.quora.com/What-is-the-difference-between-molecular-and-atomic-orbitals

Wave Functions define Energy Levels of Electrons

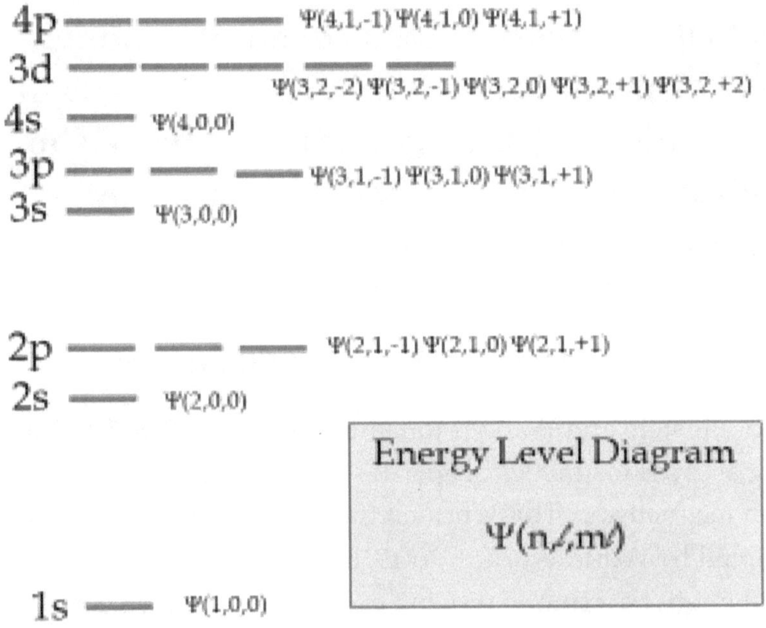

4p ▬▬ ▬▬ ▬▬ $\Psi(4,1,-1)\ \Psi(4,1,0)\ \Psi(4,1,+1)$

3d ▬▬ ▬▬ ▬▬ ▬▬ ▬▬
$\Psi(3,2,-2)\ \Psi(3,2,-1)\ \Psi(3,2,0)\ \Psi(3,2,+1)\ \Psi(3,2,+2)$

4s ▬▬ $\Psi(4,0,0)$

3p ▬▬ ▬▬ ▬▬ $\Psi(3,1,-1)\ \Psi(3,1,0)\ \Psi(3,1,+1)$

3s ▬▬ $\Psi(3,0,0)$

2p ▬▬ ▬▬ ▬▬ $\Psi(2,1,-1)\ \Psi(2,1,0)\ \Psi(2,1,+1)$

2s ▬▬ $\Psi(2,0,0)$

Energy Level Diagram

$\Psi(n,\ell,m\ell)$

1s ▬▬ $\Psi(1,0,0)$

As shown above, the principal quantum number (n) is the main determinant of energy, but because the shape affects the relative distance of the electrons from the nucleus, it also affects the energy. As we go up the energy levels, they get closer together (as shown in the Bohr model of the atom) so the 4s orbital actually has lower energy than the 3d orbitals. Obviously, as electrons are added to these orbitals, they start in the lowest energy level near the nucleus (Coulomb's law) and then move out to higher energy levels.

The electrons in the outermost energy levels are the ones that are involved with chemical bonding. They are called the *valance electrons* and elements in the vertical columns of the periodic table have similar chemical properties because they have similar valence electron configurations.

The Magnetic Moment of the Electron

In 1915, Einstein and Wander Johannes de Haas (1878–1960) were looking at gross magnetics properties of matter that hinted at intrinsic electron magnetism. This work on the Einstein-De Hass effect was interrupted by World War I. [40] In 1916, A. L. Parson suggested that the electron might be a ring of moving current generating a corresponding magnetic field.[41] But the issue of magnetism was obscured by the revolution in thinking caused by the Rutherford-Bohr model of the atom (discussed above).

---

[40] A. Einstein, W. J. de Haas, 1915, Experimental Proof of Ampère's Molecular Currents, *Deutsche Physikalische Gesellschaft*, 17: 152-170.

[41] Magneton: Theory of the Structure of the Atom, *Smithsonian Misc. Collection*, 1916, 80 pp.

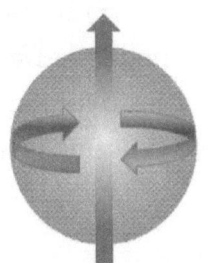

In 1921, Arthur H. Compton (1892–1962) pointed out that a spherical charge spinning on its axis would also induce a magnetic moment.[42] This hypothesis gave rise to the idea of an inherent *electronic spin magnetic moment*, which is still the model used today.[43]

The issue was clouded because in the Bohr model, an electron was assumed to circle the nucleus of an atom. This movement of a charged particle should cause an *orbital magnetic moment* (like any other circuit). Otto Stern (1888–1969) and Walter Gerlach (1889-1979) attempted to measure the magnetic moment of silver ions deflected by a magnetic field. Their intent was to measure the orbital magnetic moment. Indeed, they did see a deflection of silver ions passing through a magnetic field. Some of the atoms were attracted and some were repelled. Suggesting that the magnetic moment of the ion was aligned either with the field or against the field. It turned out that the magnetic moment they were observing was not the *orbit moment*, but rather an inherent

---

[42] A.H. Compton, The magnetic electron, *J. Franklin Inst.* 192, 144 (1921).

[43] I personally envision the electron as a photon that is stationary: It still has its magnetic and electric vectors, but the momentum has been conserved as mass (energy) as required by Einstein's equation.

electron spin moment. In 1925, George Uhlenbeck (1900-1988) and Samuel Goudsmit (1902-1978) formalized the "hypothesis of the magnetic electron" and simultaneously explained the effect of splitting the lines in the atomic spectra (observed by Zeeman in 1897) and the deflection of the silver ions by a magnetic field observed by Stern and Gerlach.[44]

Wolfgang Pauli[45] used quantum mechanical arguments to explain electron spin and from this concluded that *no more than two electrons can occupy a small region of space.* This he called the "exclusion principle" and for it he won a Nobel Prize. But you do not need to invoke quantomechanics to explain this effect. Recall that an orbital is a defined volume of space. Clearly one electron can exist there. The second electron, however, can only enter the orbital if the magnetic moments of the electrons are "paired" (in opposition, ↑↓) such that the electrostatic repulsion is balanced by the magnetic attraction. Now, realize that if you have an electron pair in an orbital and try to place a third electron into the same space, there can be no compensation of electrostatic repulsion by magnetic attraction:

↑↓ + ↑ → **electrostatic repulsion with no magnetic attraction**

Thus, you cannot place more than two electrons in any orbital and to the extent possible you will keep electrons in different

---

[44] Goudsmit had worked with Zeeman at the University of Leyden. The young students published two papers and they apologetically cited Compton in the second paper.

[45] In my opinion, a thoroughly over-rated scientist.

orbital (i.e., different regions of space) to minimize the electrostatic repulsions (chemists call that "Hund's rule").

Remember that *the electrons of the atoms determine the chemistry of the element.* Watch what happens when we build these atoms!

2.7 Electron Configuration Explains the Periodic Chart

Electron Configuration and the Periodic Chart

Notice that every element has a new *electron configuration* (arrangement of electrons).

Hydrogen $1s^1$

Helium $1s^2$ <u>First principal quantum number filled</u>

Lithium $1s^2\, 2s^1$

Beryllium $1s^2\, 2s^2$

Boron $1s^2\, 2s^2\, 2p^1$

Carbon $1s^2\, 2s^2\, 2p^2$

Nitrogen $1s^2\, 2s^2\, 2p^3$

Oxygen $1s^2\, 2s^2\, 2p^4$

Fluorine $1s^2\, 2s^2\, 2p^5$

Neon $1s^2\, 2s^2\, 2p^6$ <u>2nd principal quantum number filled</u>

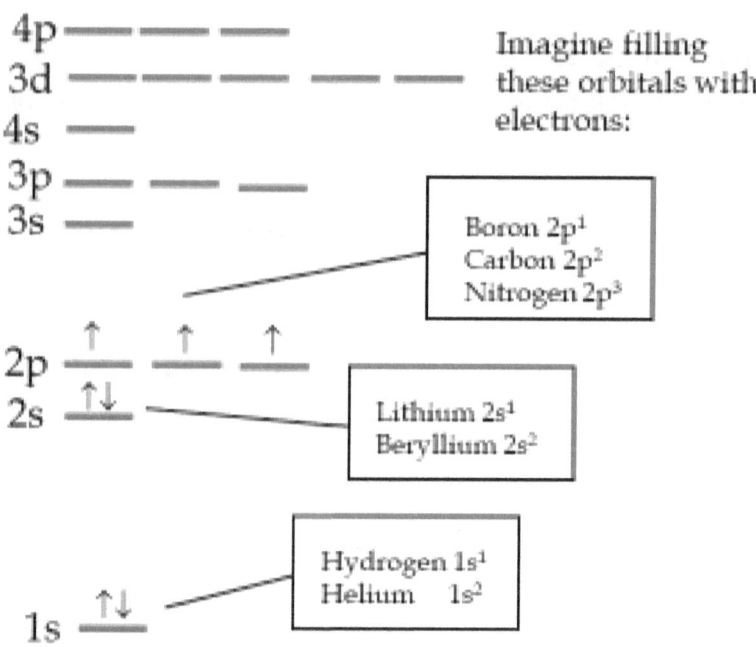

It is time to look at the modern Periodic Chart: The electron configuration of the first 20 elements are shown below.

Note that in this section, we are assuming that the atoms are in free space. When they are surrounded by ions or dipoles which create non-spherical electric fields, the atomic orbitals can be stabilized or destabilized. This is particularly interesting with the d-orbitals and results in "splitting" of the d-orbitals into high energy and low energy groups depending on the symmetry of the external field.

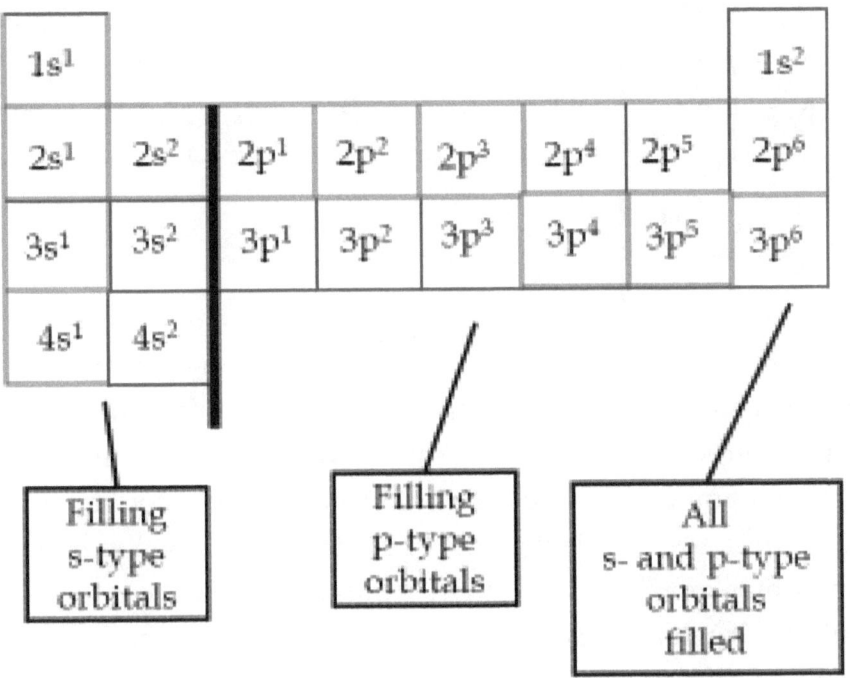

You should note that the ROWS correspond to the principal quantum number (n = 1, 2, 3, 4…etc.). In COLUMNS the elements have similar valance electron configuration and present similar chemistry (e.g., reactions). All the elements in the first column (on the left) readily lose one electron and except for hydrogen are normally metals that form ionic compounds.

H								He
Li	Be	B	C	N	O	F		Ne
Na	Mg	Al	Si	P	S	Cl		Ar
K	Ca							

Hydrogen is a special case because it only has one electron; and if that electron is removed, the only thing that is left is the nucleus. As discovered by Rutherford et al. the nucleus is very small (only ~10^{-15} m) compared to the orbits of electrons (i.e., atomic radii ~10^{-10} m). Thus, the electric field near the hydrogen ion (H+) is many times more powerful than around any other ion. As a result, hydrogen forms a diatomic molecule rather than a metal (as will be explained shortly).

The elements in the second row from the left (Be, Mg, Ca...) are metals that typically give up 2 electrons. Similarly, boron and aluminum (B, Al...) typically give up three electrons. But keep in mind that as the charges on small ions increase, releasing electrons (or gaining electrons) becomes less desirable.

 This brings us to carbon and silicon (C, Si ...). They could conceivably gain or lose 4 electrons, but forming such ions is unlikely. Thus, as in the case of hydrogen, they tend to form non-ionic allotropes and compounds (discussed below).

Looking to the right-hand column (He, Ne, Ar...), these elements were some of the last to be discovered because they are unreactive and are only found in the gas phase (they are called the inert gases). Their lack of reactivity is immediately obvious from the electron configuration: There are no empty orbital in each principal quantum number (row). So, accepting electrons, would require the electron to go into a much higher energy orbital. Moreover, there is no advantage of losing electrons, because the number of electrons they have is perfectly balanced by the nuclear charge. These elements are thus uniquely monatomic allotropes and there is not even enough interaction among the atoms to cause them to liquify. If you were going to model a helium atom, think of it as rigid as a billiard ball and as and as light as a balloon filled with air...but very small. You could consider the elements below helium as larger, heavier and softer.

To the left of the inert gases, there is a family of elements missing a single electron in their outer orbitals. These elements (F, Cl, Br...) can acquire and electron to pair with their single electron and in doing so, they achieve an electron configuration identical to the inert gas to their right. It is worth noting that when (for example) sodium (Na) loses an electron, it acquires an electron configuration identical to neon (Ne); and when fluorine gains and electron, it also achieves an electron configuration identical to neon. Thus, it is easy to understand how these reactions can occur to form ionic salts:

$$Na + F \rightarrow (Na+) (F-)$$

Of course, you are undoubtedly more familiar with table salt (NaCl).

Following the same logic, oxygen and sulfur (O, S ....) can gain two electron and nitrogen and phosphorus (N, P...) can gain three electrons. But as discussed above, forming highly charged ions is not the most favorable option for these elements. Both nitrogen and oxygen form diatomic molecules as their favored allotropes. Even the F, Cl, Br... family (i.e., the halogens) favors diatomic allotropes.

Now, we must say a few words about the transition metals. This is the block of elements that disrupts the periodic chart after calcium (Ca, see above). Looking back at the energy level diagram, it turns out that the 4s orbital is a lower energy than the five 3d orbitals. Thus, after calcium (Ca), we begin filling the 3d orbitals (following Hund's rule): First one electron is placed in each of the five 3d orbitals. That results in numerous unpaired electrons and accounts to the very strong magnetic moment of manganese:

Sc	Ti	V	Cr	Mn	Fe	Co	Ni	Cu	Zn
d^1	d^2	d^3	d^4	d^5	d^6	d^7	d^8	d^9	d^{10}
1	2	3	4	5	4	3	2	1	0

Maximum Unpaired d Electrons

Note that although the 4s orbital is lower energy in the isolated atom and are filled first, when electrons are taken away. they are typically taken from the highest principal quantum number. Thus, when the first-row transition metals (Sc...Zn) lose electrons, they normally lose the 4s electrons *before any* 3d electrons. As a result, all of the first-row transition metals form 2+ ions, by loss of their 4s electron. It is noted that copper (Cu) can form a 1+ ion, in which the second electron actually goes into the 3d orbitals to form a $Cu+/3d^{10}$ ion.

As mentioned above the d-orbitals in actual compounds are frequently exposed to external fields. Imagine that a spherical negatively charged field closed in around electrons in the d-orbitals of a transition metal. Of course, the energy of all the orbitals will be raised relative to the atom is free space. Now, suppose the same net charge was clumped together into six charges that were arranged around the ion on the principal axis. Those d-orbitals that point directly along the principal axis (dz^2 and dx^2-y^2) will be raised, while those d-orbitals that point between the principle axis (dxy, dxz and dyz) will be lowered (relative to the spherical field). It turns out that the energy level separation in many transition metals is in the range necessary to transitions to produce absorptions in the visible range. Hence, this phenomenon accounts for the colors of many transition metal ions in crystals or in solution (where the solvating molecules) provide non-spherical fields.

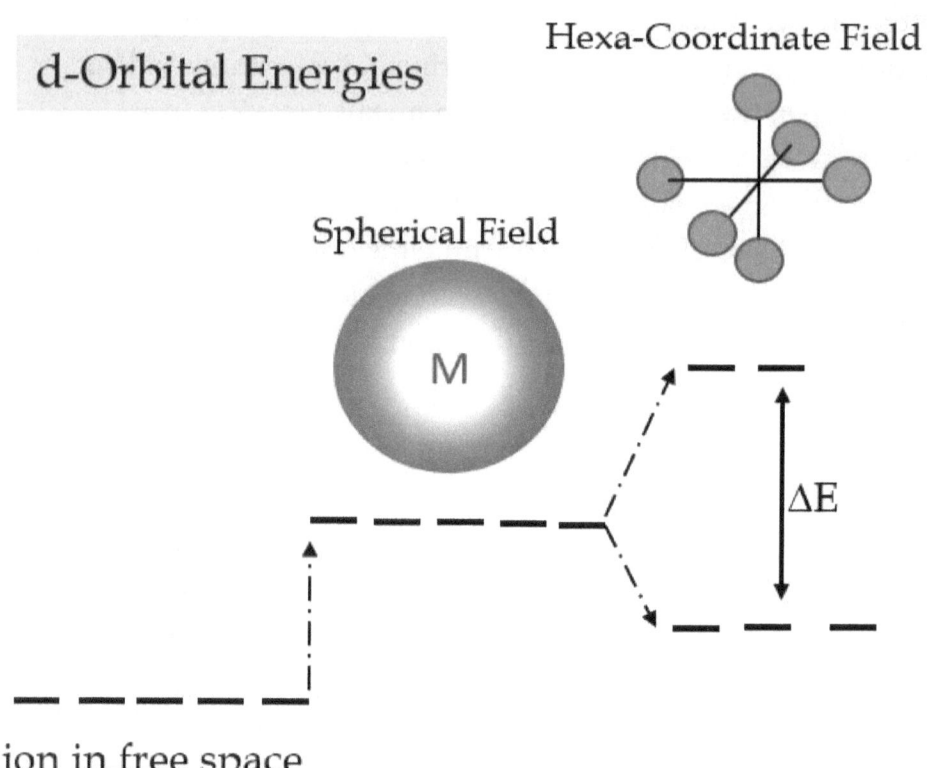

d-Orbital Energies

Hexa-Coordinate Field

Spherical Field

M

ion in free space

ΔE

It should also be noted that if the energy levels are not very far apart, they may have unpaired electrons in the upper and lower sets of orbitals (many unpaired electrons and high magnetic moment). But if the energy levels separation is large, electrons will preferably pair up in the lower energy levels (few unpaired electrons, low magnetic moment).

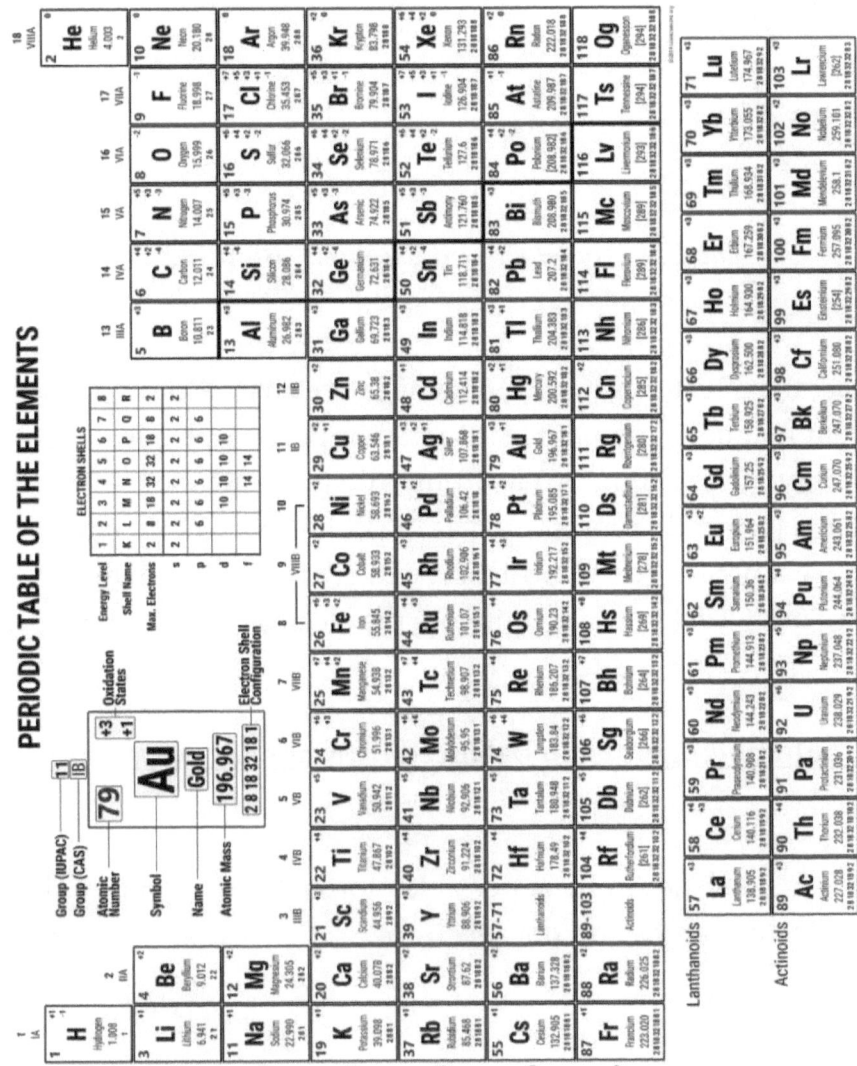

Source: https://sciencenotes.org/printable-periodic-table/

3.0 Beyond the Ionic Bond

3.1 Covalent Bonding

The Electron Pair and Covalent Bonding

The fact that electrons have both a negative charge and a strong inherent magnetic moment leads to an important physical possibility: By pairing the electrons with opposite magnetic moments (typically represented as opposite spins ↑↓) the pair will have a net magnetic attraction which at certain ranges can balance or at least partially mitigate the electrostatic repulsion[46]:

__Coulombic Repulsion__ balanced by __Magnetic Attraction__

Let us consider the simplest possible molecular ion composed of two positive charges and one electron:

---

[46]Thus, it seems that it would be possible to calculate a radius (r) at which two electrons could exist as a stable electron pair. Based on the dimensions of chemical bonds this radius must be less than 10^{-10} m.

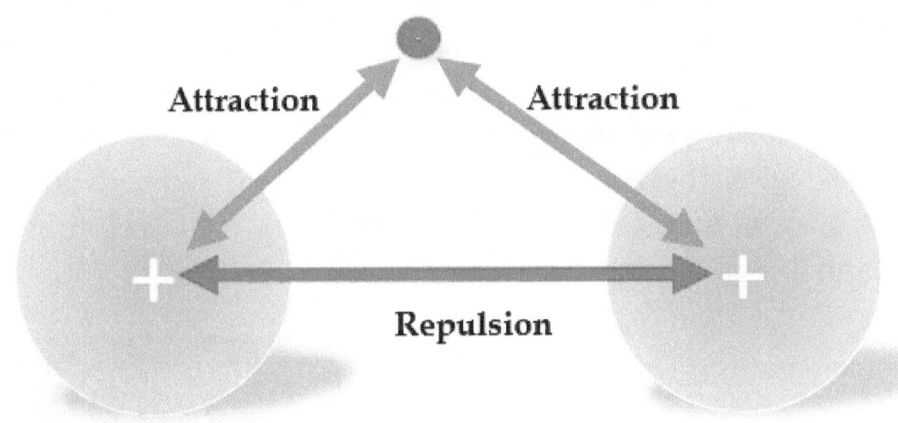

Although the electron is attracted to both nuclei, there is a repulsion between the positively charged nuclei. Now consider a pair of nuclei with two electrons:

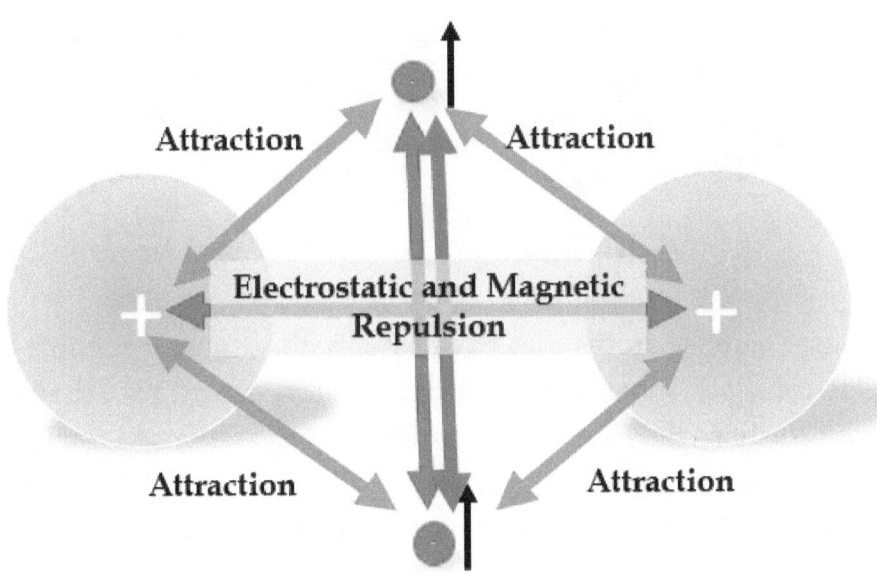

If the electrons are both in the same spin state (↑↑) they will have both electrostatic and magnetic repulsion as shown above. But if the magnetic moments of the electrons are coupled to form an electron pair, the two negative charges can be concentrated in a small region of space, which provides a powerful glue to hold the positive charges together as shown below (i.e., an electron-pair bond).

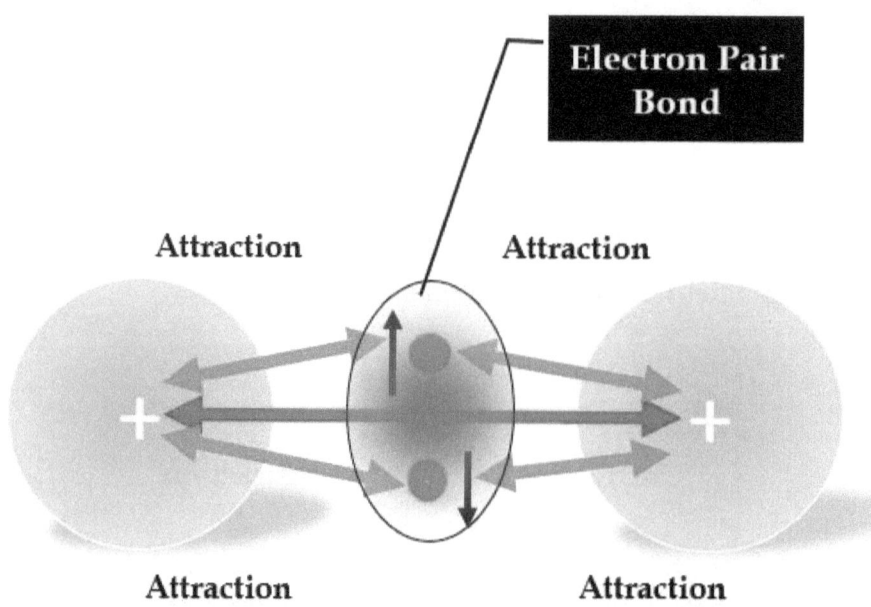

Thus, for those atoms that have essentially equal affinity for the electrons (such that formation of ions is not favorable), they will tend to share the electrons in *covalent bonds* composed of electron pairs. The electron pair is located in a hybrid orbital produced by the overlap of atomic orbitals from the two atoms.

3.2 Linus Pauling and Structural Chemistry

Optical Activity

Until the work of Schrodinger (1926), there was no physical basis for assigning any sort of directionality to chemical bonds. Ironically, chemists (especially organic chemists) had been drawing three-dimensional structures of molecules that violated the ideas of Kolbe and the electrostatic theory of bonding at least since the time of Louis Pasteur (1848) and August Kekulé (1865).

It had been observed that beams of light have yet another property: They can be polarized. Polarization can most easily be interpreted if the cyclic pulsation of the electric and magnetic vectors of light (which give light its wave properties) are interpreted as the projection of vectors of constant magnitude onto two perpendicular planes as the vectors rotate around the line of travel of the photon. Normal light (generated by random events) has equal numbers of photons rotating in each direction with no net polarization. But if the light reflects off of a surface or is passed through a material that selectively absorbs one of the rotations, the number of photons of each rotation-type will be imbalanced and the light will have a net polarization.

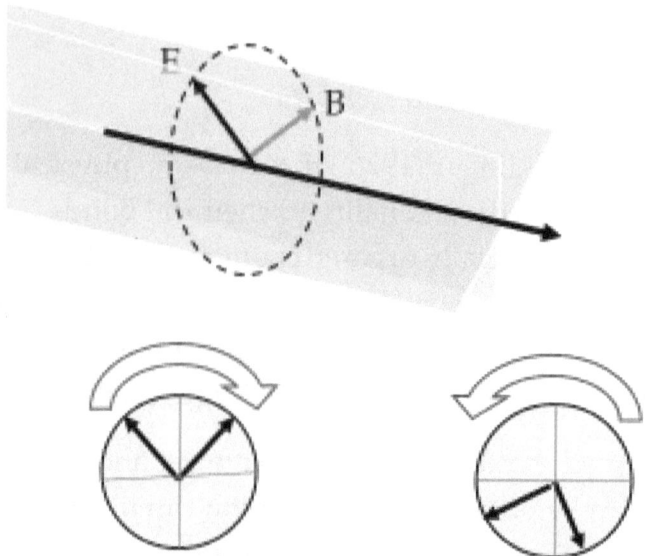

The vectors rotate in either direction
and as they rotate
their projections in the planes cycle

It turns out that some molecules have mirror images that do not interconvert and do not superimpose. Look at your hands, they are mirror images, but they do not superimpose. The most common situation is found anytime four different substituents are bound to carbon, which creates left- or right-handed electromagnetic fields around the central carbon atom. These fields interact differently with the left and right rotating photons (i.e., different absorption). We call this phenomenon *optical activity* and it is a common phenomenon in carbon-based molecules.

Pauling and Hybridization

Needless to say, organic chemists (focusing on molecules based on carbon) were not happy with Schrodinger's simple atomic s, p and d orbitals, which did not explain the shapes of organic molecules. These shapes were being proven to be real by x-ray crystallography by 1928.[47]

Fortunately, to accurately reflect the three-dimensional distribution of electrons in most molecules, the atomic wave functions can be combined in various linear combinations:

$$\Psi_{molecular} = \Sigma\ \Psi_{atomic}$$

This idea was added by Linus Pauling (1901–1994) who took Schrödinger's work from isolated atoms to real molecules in his book *Nature of the Chemical Bond* (1939).[48]

For example, using s- and p-type wave functions, bonding *hybrid atomic orbitals* can be constructed as follows:

$\Psi s \pm \Psi p$ → two hybrid orbitals at 180° angle

$\Psi s \pm 2\ \Psi p$ → three hybrid orbitals at 120° angles

---

[47] Cox, E. The Crystalline Structure of Benzene. *Nature* 122, 401 (1928).

[48] Nobel Prize in Chemistry 1954 "for his research into the nature of the chemical bond and its application to the elucidation of the structure of complex substances."

$\Psi s \pm 3\ \Psi p \rightarrow$ four hybrid orbitals at 109° angles

These combinations (sp, sp², sp³) account for most of the shapes found in molecules made of C, H, O, N and halogens (linear, planar trigonal, tetrahedral).

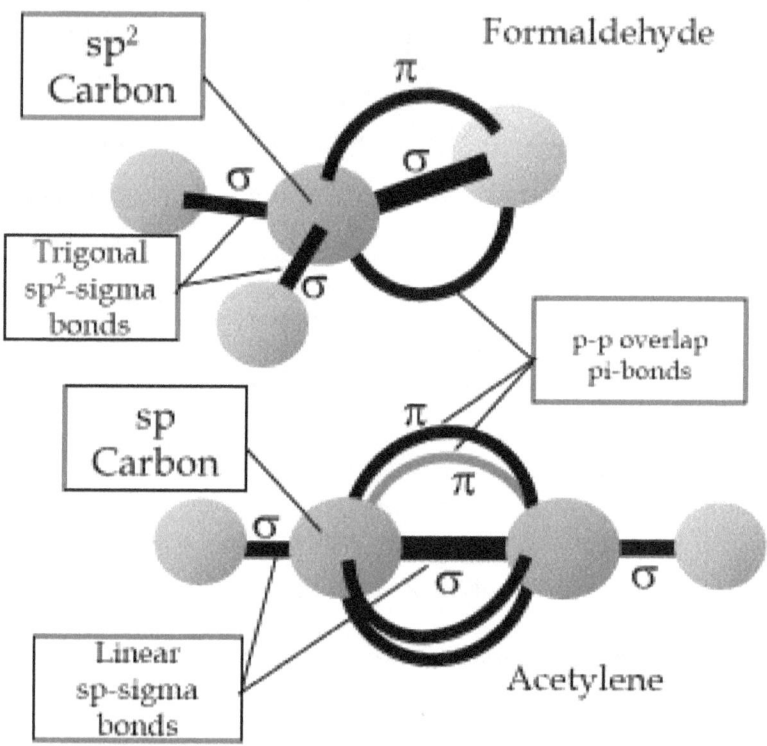

Above, examples of sp² hybridized carbon (formaldehyde $H_2C=O$) and sp hybridized carbon (acetylene HCCH) are shown. The direct overlap between atomic orbitals are called *sigma bonds* and the overlap of unhybridized p-orbitals are called *pi bonds*.

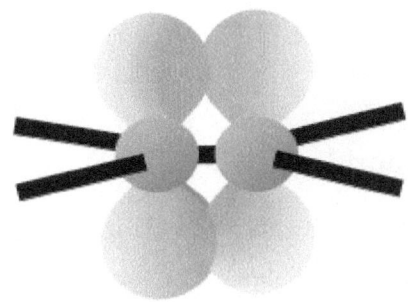

This is <u>one</u> pi (π) bond
between two sp² hybridized
carbon atoms

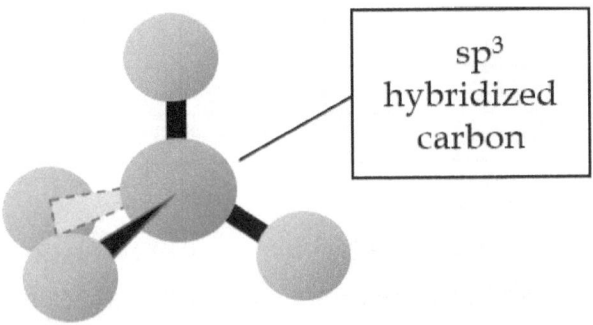

sp³
hybridized
carbon

Tetrahedral
bond angle 109°

The Metals

So far, we have explained the ionic bonding of salts and the covalent bonding of molecules; but neither of these explains the behavior of metals. Compare the properties of sodium metal (Na) and sodium fluoride (NaF).

First, we note that sodium metal readily conducts electricity, but sodium fluoride dos not (unless it is melted or dissolved in water). This indicates that in sodium fluoride there are no free electrons and the ions are not free to move. On the other hand, there must be free electrons in sodium metal.

Second, compare the melting and boiling points of sodium and sodium fluoride:

	Melting Point (°C)	Boiling Point (°C)
Sodium Metal	97.8	883
Sodium Fluoride	993	1,704

The electrostatic forces holding sodium fluoride together are much stronger than the forces holding sodium metal together.

Third, we know the relative sizes of the Na^+ and F^- ions (which are isoelectronic, i.e., they both have exactly 10 electrons) and we know the crystal structure of NaF and we could calculate the energy holding the crystal together using Coulomb's law (if we had a powerful computer). We call this energy the "lattice energy" (see above).

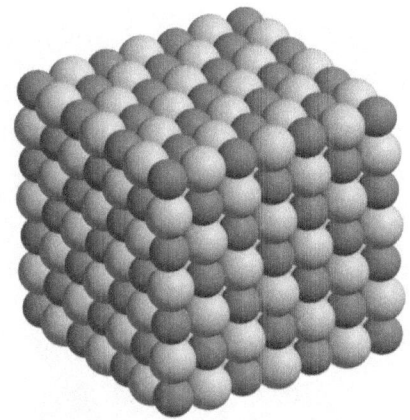

Author: Benjah-bmm27; Source Wikimedia Commons

Compare the atomic mass of sodium and fluorine:

Sodium: 22.99

Fluorine: 19.00

Thus, if the atoms in sodium metal were packed the same way the ions are packed in NaF, sodium fluoride would be slightly less dense than sodium. But here is the big surprise:

	Density (g/cm³)
Sodium Metal	0.968
Sodium Fluoride	2.558

Sodium fluoride is more than twice as dense as sodium metal! Imagine that in the crystal structure above, all the spheres of one color disappeared and the other atoms actually moved apart a little. The *ionic radius* of sodium (Na+) is 0.116 nm (1.16 Å), and the *atomic radius* of sodium is 0.154 nm. Basically, the electrons in the outermost orbitals of the sodium atoms in metallic sodium form a field of negative charge in which the core Na^+ ions float. This model explains why sodium is malleable and can even be cut with a dull knife and why the electrons are free to move under an electric potential.

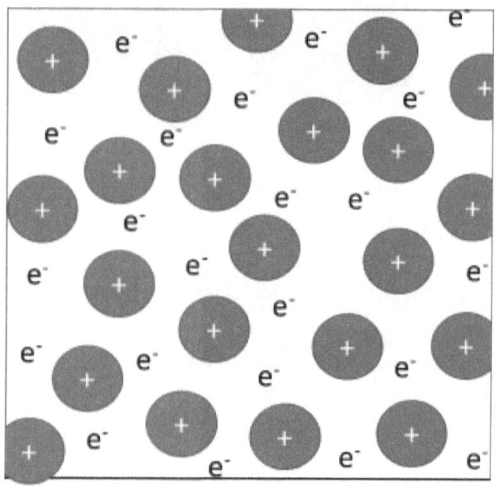

When you hit a metal with a hammer it deforms because the atomic cores can slide past one another. When you hit a crystal with a hammer it shatters because there is no way for the crystal to deform without breaking bonds.

3.3 Chemical Reactions

Breaking and Making Bonds

Atoms, molecules and ions undergo chemical changes by breaking and making chemical bonds. In some cases, it is feasible to break a bond completely before making a new bond. However, in most cases this sort of *mechanism* would usually requires a very large activation energy (Ea, see section 1.2). The highest point in the pathway is called the *transition state*.

To minimize this unfavorable situation, bond making is often concurrent with bond breaking. The term that chemists use is "concerted," everything is happening at once, as opposed to "step-wise."

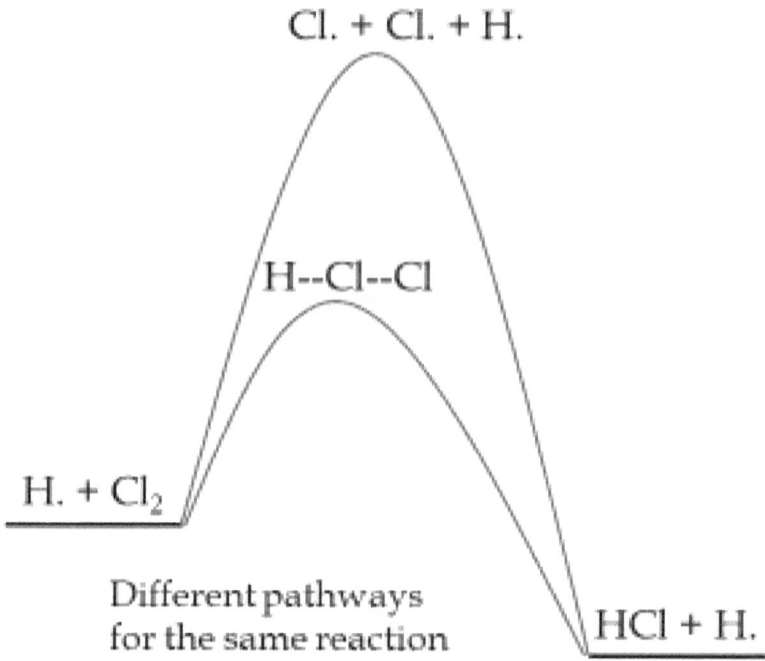

Cl. + Cl. + H.

H--Cl--Cl

H. + Cl$_2$

Different pathways
for the same reaction

HCl + H.

4.0 Organic Chemistry: A Brief Introduction

Organic chemistry is a branch of chemistry devoted to
compounds of carbon. This element has become the focus of
enormous attention because of the fact that it forms strong
covalent bonds with itself and a variety of other elements
especially H, N and O but also halogens, S and P. There was a
time when it was believed that only living systems made these

compounds, but now it is possible (not necessarily practical or economical) for chemists to make almost any compound made in nature.

4.1 Hybridization and Molecular Shape

Carbon, hydrogen, nitrogen, oxygen, etc. allow the formation of a wide variety (make that "almost limitless") number of shapes of chemical compounds. Carbon, nitrogen and oxygen form bonds with essentially three different hybridizations of their atomic orbitals:

sp-hybridization produces two bonds pointed in opposite directions along one axis, leaving two unhybridized p-orbitals on perpendicular axes (as shown in acetylene, above).

sp^2- hybridization produces three orbitals in a plane with an angle of 120-angular degrees between them. The lone unhybridized p-orbital is perpendicular to that plane (e.g., consider the formaldehyde conformation above and below).

sp^3-hybridization produces four orbitals pointed at the corners of a tetrahedron (109.5 angular degrees apart) with no unhybridized p-orbitals.

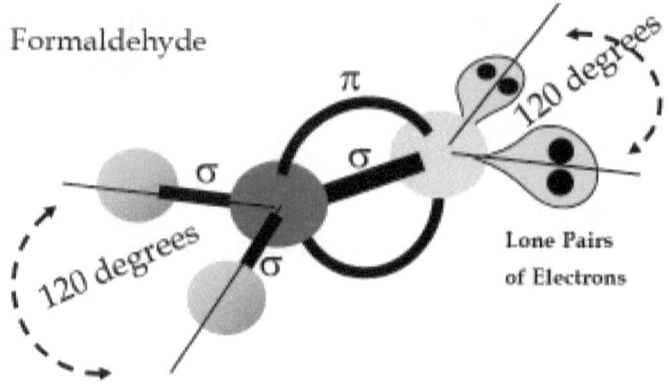

Formaldehyde

Both the Carbon and Oxygen
are sp² Hybridized

The Oxygen Atom has two Lone Pair
of Electrons

Each orbital can contain 2 electrons. Those electrons are either provided from the atom itself (as "lone pairs" of electrons); or one electron can come from each atom forming a *covalent bond*; or an empty orbital can actually be filled by donations of a lone pair from another atom forming a *coordinate covalent bond*.

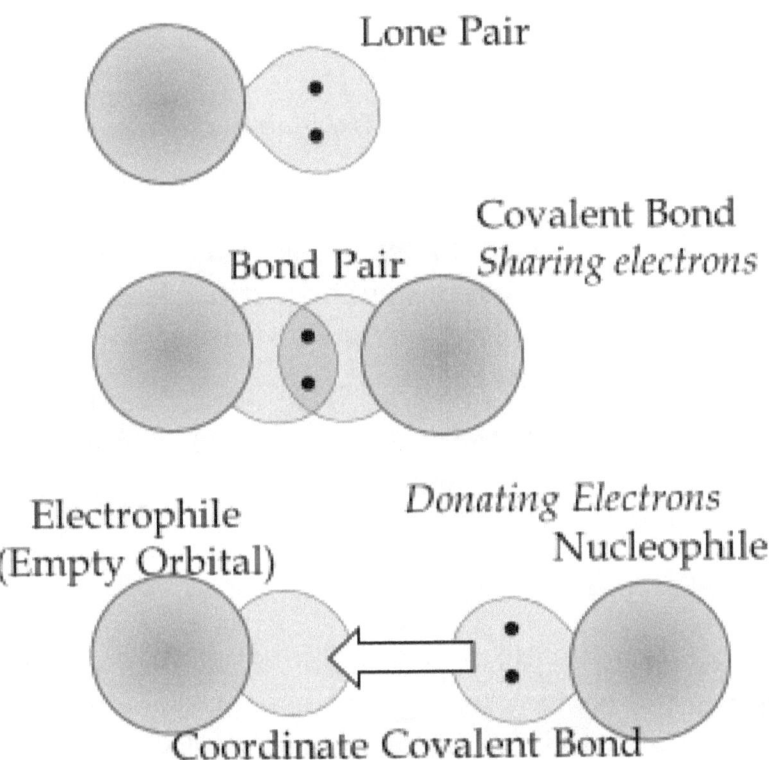

Atoms that have empty orbitals or "electron deficient" orbitals are called electrophiles (they want electrons) and atoms that have lone pairs of electrons are nucleophiles (seeking positive charge).

If you understand these shapes, you can quickly predict the shape of most organic compounds and likely reactions.

4.2 Structure and Reactivity

Organic Chemistry Shorthand

Because of the complexity of many organic molecules, chemists tend to draw diagrams to represent their structures. This may seem confusing at first; but it is actually very easily to understand. Here are the rules:

(1) Each angle or end of a line represents a carbon atom (if another atom is intended, it will be shown by its symbol (e.g., N or O);

(2) each carbon has four bonds;

(3) if the bonds shown do not account for all the carbon bonds, then the carbons are assumed to have hydrogen(s) attached or will show a positive charge or the number or electrons in the orbital (. or :);

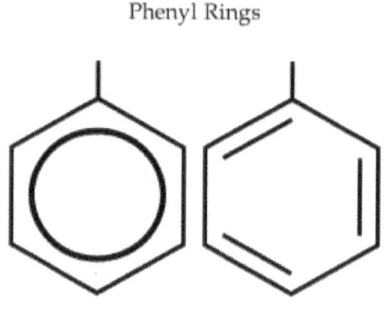

Phenyl Rings

(4) multiple bonds are indicated by multiple lines (= is a double bond);

(5) the benzene ring (phenyl group), may be represented with bonds or a ring representing the delocalized cloud of electrons in unhybridized p-orbitals.

Reactivity of Organic Compounds

Notice that the unhybridized p-orbitals of the sp^2 hybridized carbon atoms of the benzene ring are exposed above and below the ring and they all overlap into six hybrid orbitals represented by the circle in the representation (above).

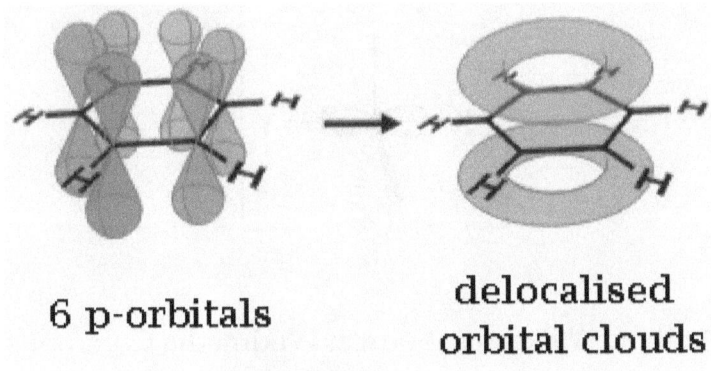

6 p-orbitals　　　　delocalised orbital clouds

Author: Apalmer; Source: www.wikidoc.org

These rings of electron density are nucleophilic and react with electrophiles (such as NO_2^+, Br^+, $H_2C=O$), which can displace the hydrogens making benzene a very important starting material for synthesizing new compounds.

Most sp^3 hybridized carbons with hydrogen substituents are relatively unreactive and we are inclined to simply represent them as an undefined hydrocarbon group "R-" where R stands for "radical." For example, the methyl group (H_3C-).

However, if a "good leaving group" (i.e., a substituent group that forms a stable anion in solution such as Br-, or NO_3^-) are attached to a sp³-hybridized carbon, it can be displaced by a good nucleophile. If the leaving group is very stable and the solvent is very polar, the leaving group (Y) may dissociate on its own. But (more commonly) the in-coming nucleophile (X) assists the leaving group depart in a concerted reaction through a transition state as shown below:

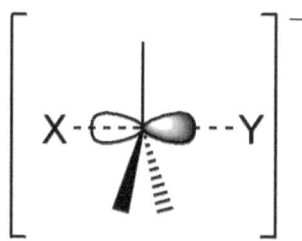

Author: Walkerma, Source: Wikimedia Commons

4.3 Nomenclature

A number of nomenclature devices (with historical roots) are used by chemists. But we do not need to go into much depth here for pure hydrocarbons.

On the other hand, certain structures (typically involving N, O, and halogens) are places were reactions commonly occur. These are called "functional groups" because they are expected to undergo certain types of reaction regardless of the overall composition and structure of the molecule (R). Here are a few of the most common:

R-X halide (X = F, Cl, Br, I)

R-OH alcohol

R-SH thiol

R-O-R ether

Compounds marked with a * contain the carbonyl (C=O) group

R-CO-H aldehyde*

R-CO-R ketone*

 R-CO-OH carboxylic acid*

R-CO-OR ester*

R-CO-NH₂ amide* (H can be replaced with R)

R-NH₂ amine (H can be replaced with R)

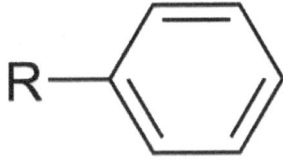 phenyl (aromatic, "Ar")

To simplify structures in drawings, organic chemists typically don't identify hydrogens bound to carbon and represent carbon as an angle in a structure or the end of a line. But in order to represent the three-dimensional structure, bonds to hydrogens and other elements may be represented by a wedge (coming out of the page) or dashes (going behind the page). In the drawing below, the solid lines lie in the plane of the paper and the bold bonds to hydrogens are coming out of the paper, while the dotted lines to hydrogens are going into the paper. The following structure incorporates a number of these short-hand features:

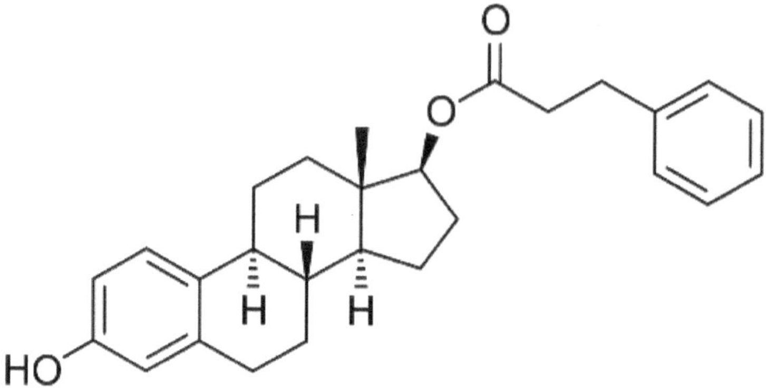

Source: Wikipedia Commons

It is important to note that these molecules are three-dimensional and chains of -CH_2- units (methylene units) are flexible (i.e., the chain can fold into various conformations by rotating around C-C single bonds).

Here are a few names that are worth remembering:

CH_4 methane H_3C- methyl H_3C-OH methanol

H_3C-CH_3 ethane H_3CH_2C- ethyl H_3CH_2C-OH ethanol

Nomenclature will be introduced from time to time in the text below, but is generally not critical for understanding.

Part II. Materials of Civilization

5.0 The World in which we Evolved and Live

5.1 Brief History of The Solar System and the Earth

In the Beginning

Humans have always tried to understand our history. Most religions include a "creation story." If these stories are read as "hypotheses" presented by thoughtful observers, we can appreciate the logic and understanding that they represent; and by testing these hypotheses, we can come to a better understanding over time. The problem occurs when the innovative thinking of open-minded philosophers is converted into a religious dogma, which is held beyond question. In western history, I view the creation hypothesis passed down by the Judaic tradition as truly inspired. Unfortunately, it became blended with Greek philosophy, which held that truth could be derived from logic and reason *alone*. What the Greeks failed to understand is that logic is "truth preserving," not truth creating: Applying logic to a *true proposition* will produce a true result. But, if you start with a *false proposition* and apply logic to it, the result may or may not be truth. Unfortunately, the results of the reasoning of Aristotle (384–322 BCE) and his contemporaries were accepted as **facts** and given political as well as religious significance as the modern western religions took shape and dominated western Europe with the Roman Empire through 410 CE. With the collapse of the Roman Empire, there was no other basis of knowledge in Europe until the late 1600s CE. Aristotle was accepted for nearly 2000 years!

Regardless, our current hypothesis (consistent with the facts we now accept) holds that about 13.5 billion years ago, an essentially instantaneous burst of energy (electromagnetic) was released into a void of space and quickly degraded to fundamental particles: Protons and electrons, which conserved the properties (energy as mass, electric vector as charge, and magnetic vector as a magnetic field) of electromagnetic radiation. These characteristics have been identified above. Hydrogen atoms assembled from the particles with the creation of matter as we know it. Potential and kinetic energy was also associated with the particles.

Note that, my interpretation of the data is slightly different than broadly accepted by physicists; but it is irrelevant to the chemistry which follows.

Gravity drew the hydrogen atoms into stars; and in the crushing pressure of the center of stars, hydrogen atoms were compacted into a smaller volume known as the *neutron*. The existence of neutrons allowed hydrogen atoms to be compressed into a variety of heavier nuclei with continuous release of energy (nuclear fusion). Ultimately, these stars exploded releasing hydrogen and a range of heavy atoms with nuclei composed of various numbers of protons, and neutrons. In a vacuum with no confining force, many of the nuclei were unstable and decayed back to stable fragments (i.e., typically a stable nucleus with electrons, fragments of nuclei (alpha particles) and free electrons). These pieces (alpha particles and electrons) reassembled over time into helium (He).

Some of these nuclei (i.e., isotopes) are indefinitely stable because the neutrons (which are inherently *unstable*[49] without crushing pressure or electrostatic restrictions) cannot successfully dissociate into protons and electrons because the electrons cannot escape the strong electrostatic fields of the highly charged nuclei.[50]   Other nuclei are not stable, but because they have half-lives of billions of years, some isotopes have survived as radioactive nuclei (e.g., of potassium, thorium and uranium) until today. But most combinations of protons and neutrons in early stars were completely unstable and caused the explosion of the massive stars in nuclear chain reaction. Indeed, there are no stable isotopes of the element technetium (Tc) and the longest-lived isotope ^{97}Tc has a half-life of only 4.21×10^6 years as compared to the half-lives of ^{40}K (1.25×10^9 years), ^{232}Th (1.40×10^{10} years) and ^{238}U (4.47×10^9 years).

---

[49] The half-life of a neutron in free space is only 881.5 ± 1.5 s.

[50] I have written extensively about this process. I suggest you refer to the appendix in my book *History of Atomic Theory* published on Amazon as a paperback in 2019.

The nuclei acquired electrons to balance the electrical charge and formed atoms. Then the atoms collided and formed stable allotropes (metallic, monatomic, diatomic, polyatomic) by sharing electrons (as described in Part I). Hydrogen still dominated the universe. And the stable (or long-lived) isotopes of elements spiraled into a hierarchy of eddies (galaxies and solar systems).

Within a solar system, the kinetic energy of the atoms was gradually dissipated as heat in the condensing stars and planets. At some point, nuclear fusion was initiated in the stars and heated the surface with release of a continuous spectrum of electromagnetic radiation that depends on the surface temperature of the star.

Blackbody Radiation

In Part I, electromagnetic radiation was discussed; and it was noted that it is a way for "nature to make change" when energy must be conserved. There are two fundamental laws governing the behavior of physical bodies: Conservation of momentum and conservation of energy. *In a collision both momentum and energy must be conserved.* Momentum is the product of mass and velocity (mv) where the velocity is a vector (it has a direction). Thus, momentum can only be conserved in movement of particles (which must represent a certain corresponding amount of kinetic energy). Energy (on the other hand) can take on many forms including electromagnetic radiation.

Consider the following example, two identical masses (mass = m) are moving towards one another with velocities +v and -2v). The momentum of the system is

$$m(+v) + m(-2v) = mv-2mv = -mv$$

After they collide the momentum of the system must still be -mv.

Now let us look at what happens to the kinetic energy of the system: The kinetic energy of the system before collision is

$$\frac{1}{2} m(+v)^2 + \frac{1}{2} m(-2v)^2 = \frac{1}{2} m[v^2 + 4v^2] = \frac{1}{2} m(5v^2)$$

But the *kinetic energy after the collision* can only be $\frac{1}{2} mv^2$ thus, the kinetic energy of the system has changed

$$\text{from } \frac{1}{2} m5v^2 \text{ to } \frac{1}{2} mv^2$$

Thus, $\frac{1}{2}m4v^2$ of kinetic energy must either be conserved as *heat or electromagnetic radiation*. This is a general outcome:

When particles collide some of the kinetic energy is converted to electromagnetic radiation.

As atoms collide on the surface of the sun (or any other body) electromagnetic radiation is released. Because the velocities of the particles follow a Boltzmann distribution, the energies of the collisions tend to be in a Boltzmann distribution (i.e., crowded to the low end of the spectrum) but because the relative velocities of collision depend upon the angle of incidence of the two particles many collisions will be a much lower velocity (approaching zero). Thus, the radiation will be a continuum with a very long tail all the way to zero energy (infinite wavelength).

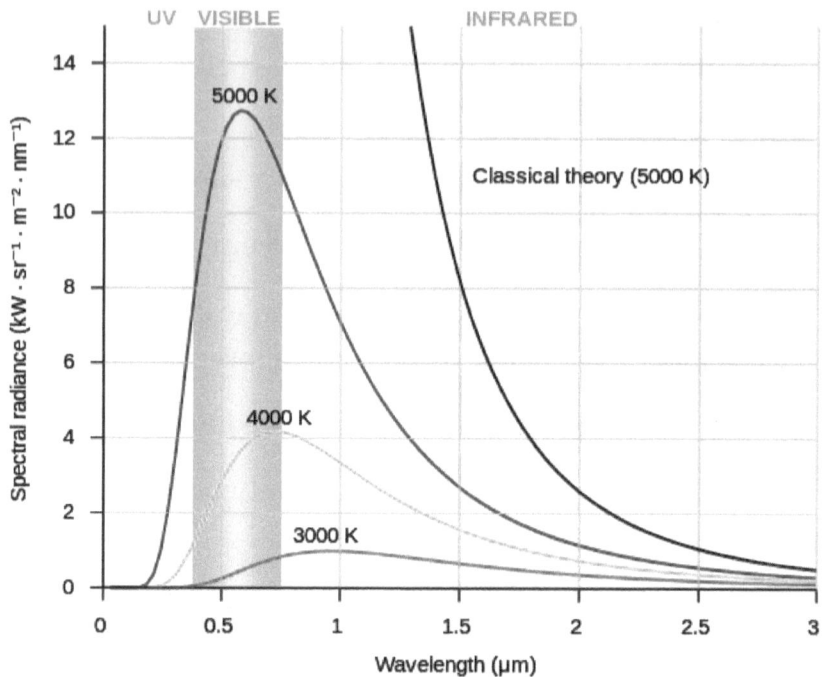

Author: Darth Kule; Source Wikimedia Commons

The Planets

Meanwhile, the planets have tended to gather around higher atomic number (denser) particles and their kinetic energy and radioactivity was converted into heat. Heat is the kinetic energy of molecules stored as vibrations, rotations and translations of molecules. Infrared radiation transfers heat to surfaces most efficiently because this is the region (750-1000 nm) of the spectrum that corresponds to the vibrations of molecules.

Because isotopes of iron are the most stable nuclei, iron was the predominant element forming the proto-planets. Kinetic energy of particles colliding ensured that the planet was molten.

Metallic iron is dense (and with other dense elements including radioactive elements) settled into the core of the planet. Other common elements (oxygen, hydrogen, aluminum, silicon phosphorus, nitrogen) form stable molten liquid or gaseous compounds. These compounds are primarily oxides or halides of silicon, aluminum and phosphorus, which crystalized into (i) a solid crust (2.7 g/mL) floating on (ii) a mantle of molten oxides and sulfides (3.3 to 6 g/mL) resting on (iii) the liquid iron core (7.9 g/mL) as the kinetic energy was dissipated.

Oxygen and nitrogen combined with hydrogen and carbon forming volatile compounds (including CO, CO_2, H_2O, NH_3, HCN, CH_4 and H_2CO) which gradually condensed into the oceans and atmosphere as the planet cooled towards 300°K on the surface. However, radioactive elements in the core (U, Th, K) generate enough heat to keep the core molten over billions of years. It is estimated that, today, the heat in the interior of the earth is approximately 20 terawatts from radio activity and 27 terawatts from "residual heat" (i.e., kinetic energy associated with the earth's spinning (47 terawatts total). This has little effect at the surface, where solar radiation provides 173,000 TW.

Meanwhile, water vapor dominates the ability of the atmosphere to pass visible light from the hot (5000°K) sun to the surface and trap longer wavelengths of light released by the cool surface of

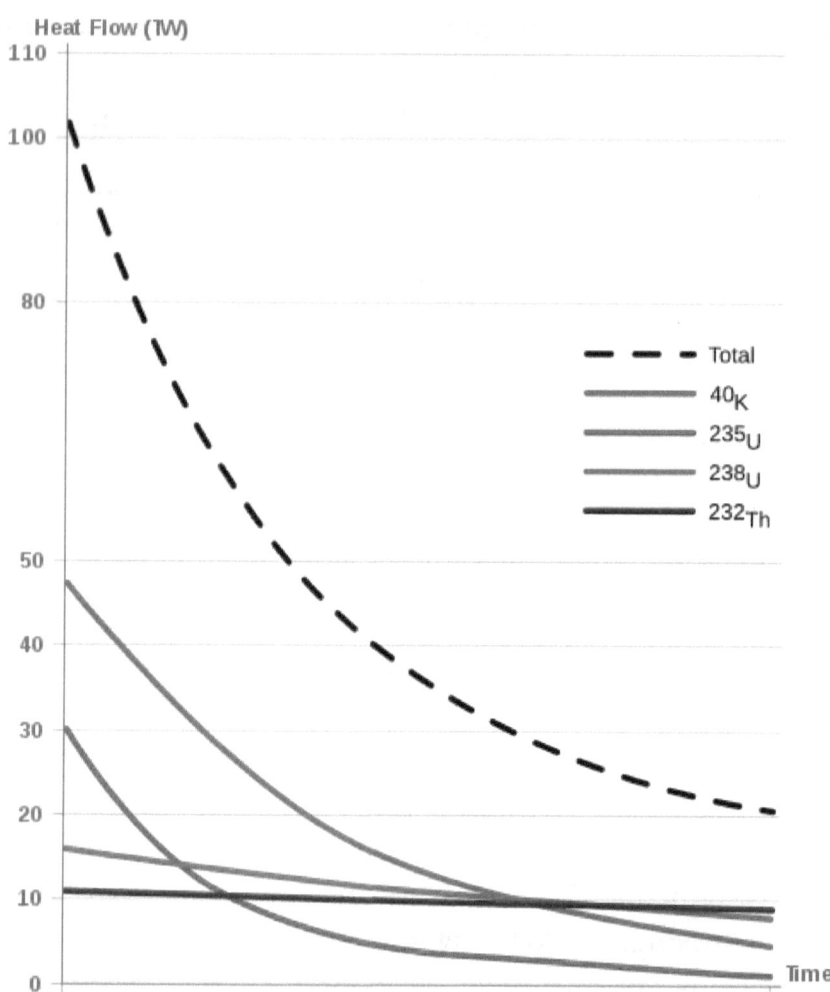

Author: Goran tek-en; Source: R8R and Wikimedia Commons

the earth. The surface temperature of the earth has not dropped below about (300°K, 0°C) because below this temperature Wien's law predicts that almost all the heat from the earth is trapped by water vapor, which absorbs strongly at wavelengths longer than 20,000 nm (20 μm).

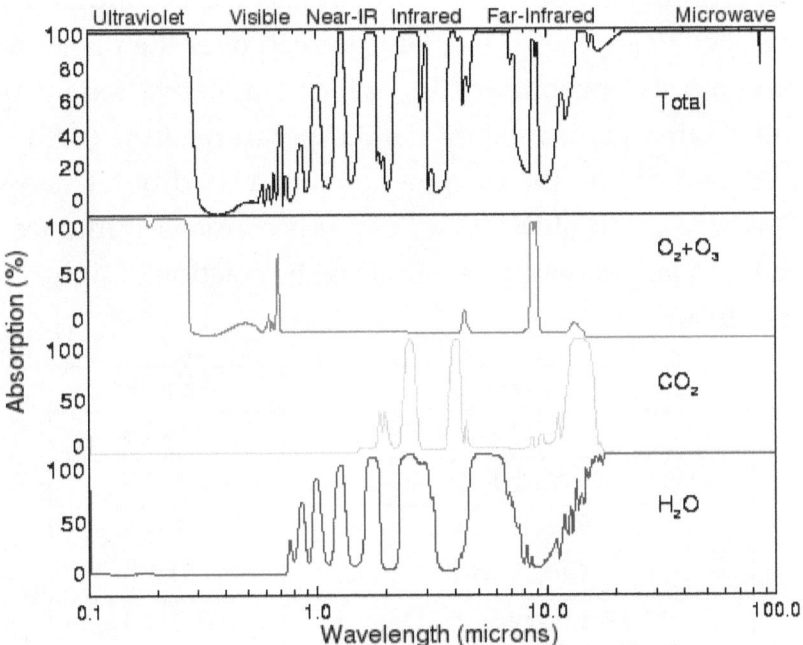

Source: Bernardjlewin@gmail.com

You will notice from absorption spectra that the "earth shine" (blackbody radiation from earth) escapes via the *spectral window* centered on 10 microns (10,000 nm). This spectral window is partially blocked by absorptions due to O_2/O_3 and CO_2.

The "global warming" debate tends to focus on the physics without much regard for the chemistry. So, I'll focus my comments primarily on chemistry:

The well-established "simple" science agrees that CO_2 will cause some global warming. This was an observation made by a chemist (Svante Arrhenius, 1859-1927) in 1896 as part of his explanation of why the earth periodically drifts into ice ages. He was hopeful that the burning of fossil fuels would mitigate the next ice age. Presumably, as plants spread over the planet and absorbed more atmospheric CO_2, which was converted into coal, peat and "soil organic matter" the temperature of the earth dropped; and did (in the extreme) go into a condition ("snowball earth') where much of the water vapor is condensed into ice (which has a large albedo, i.e. *albedo* is the fraction of light reflected from a surface).

Surface	Albedo
Fresh Snow	0.6-0.9
Cloud Tops	0.4-0.8
Old Snow	0.4-0.6
Dry Sand	0.4
Wet Sand, Dry Soil, Deserts	0.2-0.3
Crops and Meadows	0.1-0.2
Dark Wet Soil, Forest	0.05 to 0.15
Liquid water	0.08

Note that in the case of plants a significant fraction of the absorbed electromagnetic energy is converted to chemical energy through photosynthesis. Forest have multiple tiers of plants that make good use of incoming solar energy. In each encounter with a green leaf, about 25% of solar radiation is turned to chemical energy, but only about 4% is actually incorporated in carbohydrates.

High albedo has resulted in a condition known as "snowball earth where the ice sheets cover most of the planet. Recovery from this condition is presumably achieved by volcanic ash reducing the albedo and volcanic CO_2 and water vapor restoring the greenhouse effect.

However, as a chemist, it seems relevant that the combustion of coal (C) consumes oxygen from the atmosphere:

$$C + O_2 \rightarrow CO_2$$

$$\text{"}CH_2\text{"} + 2\,O_2 \rightarrow CO_2 + H_2O$$

Thus, for every mole of CO_2 that is produced, a mole of oxygen is consumed. And oxygen O_2/O_3 absorbs right in the middle of the spectral window in the water absorption spectra (centered on 10,000 nm). Although the CO_2 band is broader, it sits in a region where water already absorbs much of the exiting radiation. I could envision the reduction in oxygen and the increase of CO_2 more or less balancing out in the current atmosphere (depending on the surface temperature). Remember, Arrhenius was assuming that CO_2 could be increased *without reducing* O_2 in the atmosphere. Combustion of petroleum, takes even more oxygen out of the atmosphere per unit of CO_2 formed (the water that is produced is already saturated and presumably precipitates from the atmosphere).

Before there was plant and animal life there was no carbonate rock and no coal or petroleum. Yes, almost all of that carbonate rock was produced _after the earth cooled_ to near ambient temperature!

The White (chalk, $CaCO_3$) Cliffs of Dover

crumbling into the English Channel

The Daily Mail March 14, 2012.

Virtually all the carbon dioxide represented by all existent fossil fuels *and limestone rock*[51] were in the atmosphere…and yet the earth cooled!

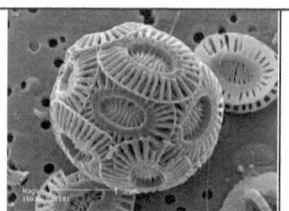	Electron microgram of chalk Produced by algae *Coccolithophores* Source: https://factsman.tumblr.com/tagged/chalk

---

[51] Note that carbonate rock is not even stable at moderate temperatures.

Perhaps, we should be more concerned about the manufacture of concrete in lime kilns:

$$CaCO_3 \rightarrow CO_2 + CaO$$

In this case, CO_2 is released without absorbing oxygen.

Equilibrium pressure of CO_2 over $CaCO_3$

(P) versus temperature (T)

P (kPa)	0.0055[52]	9.3	24	34	51	72	80	91	101	179
T (°C)	550	748	800	830	852	871	881	891	898	937

Source: Wikipedia "Calcium Carbonate"

There have been many ice ages and apparently two episodes of "snowball earth." But there is no evidence of ever having runaway heating. This is odd since the earth started out hot from kinetic energy and the radioactive decay is now the lowest it has ever been and is declining (very slowly, see above).

The frightening projections of runaway heating caused by CO_2 invoke increased atmospheric concentrations of water vapor. But that begs the question of how the earth cooled from much higher

---

[52] The current pressure of CO_2 in the atmosphere is about 400 ppm or 0.0004 atm, 1 atm = 101 kPa; thus, the current concentration of CO_2 is 0.04 kPa.

temperatures to the ambient temperature today since at higher temperature both of these components would have been at higher concentration in the atmosphere. The evaporation and condensation cycle of course transports heat at the surface to the top of the troposphere where it is released as water vapor becomes water droplets. And the clouds provide increased albedo (i.e., reflectivity of the earth).

Finally, it is odd that governments advocate limiting CO_2 emissions when H_2O is the presumed threat of runaway heating. If you do some calculations you will find that the rate of water loss from irrigation and lakes created in arid climates (e.g., Lake Nasser in Egypt or Lake Mead in the US) release more water vapor than burning coal released CO_2. So, why wouldn't governments focus on regulating (e.g., center pivot) irrigation before worrying about burning coal or petroleum. After all, plants that produce our food through photosynthesis are essentially starved for carbon dioxide.

Diffusion of Gases and Escape Velocity

In the gaseous state, the forces of attraction among molecules is less than the forces associated with changing their momentum when they collide. The random collisions result in a randomization of kinetic energy of individual molecules. We measure the kinetic energy of a gas, by its absolute temperature (°K).

If we compare molecules of different masses that are at the same temperature (i.e., the average kinetic energy of molecules at the same temperature is the same regardless of their masses). Then we can write

$$\frac{1}{2}\, m_i V_i^2 = \frac{1}{2}\, m_j V_j^2$$

If this argument is true, then we can rearrange the equation to this

$$V_i/V_j = (m_j/m_i)^{1/2}$$

This relationship is called *Grahams law* and basically says that at a particular temperature, on average small molecules will move substantially faster than large molecules.

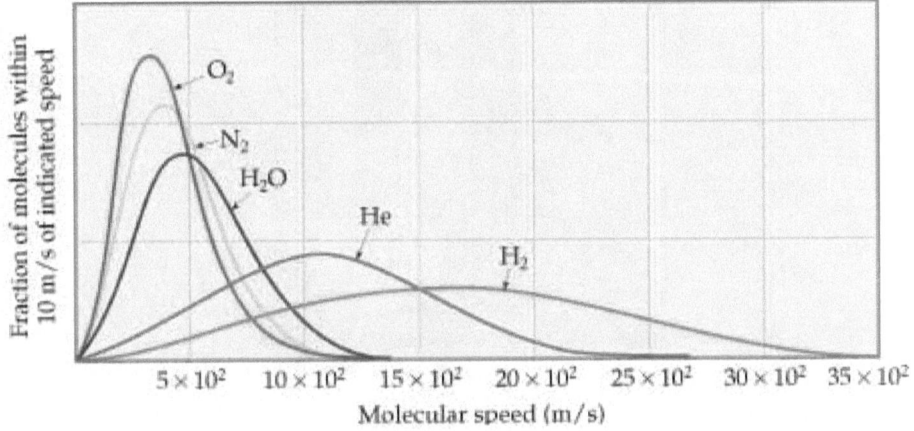

Source: Prentice Hall
http://wps.prenhall.com/wps/media/objects/3311/3391331/blb1008.html

Based on the mass (gravity) of a planet, it is possible to calculate the minimum velocity needed for a body (particle) to escape. For the earth that number is 1.12×10^2 m/s. You will notice that at

298°K (25°C), lighter molecules far exceed that and were readily lost from the upper atmosphere while the earth was hot.[53] Water and water-soluble compounds (ammonia, formaldehyde, HCN and carbon dioxide) were retained in the earth's atmosphere, however, because they condense into droplets of liquids above the surface of the earth and were constantly recycled between the hot surface and the cooler upper reaches of the atmosphere (troposphere) where they radiate heat upon condensing.

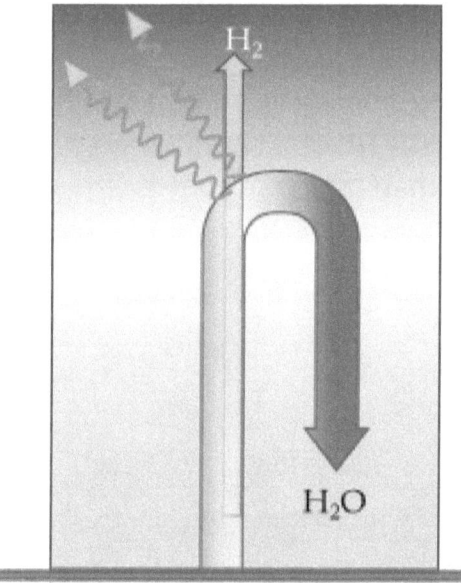

During the hot phase of the solar system, light gases (hydrogen, helium) were almost totally lost from the inner planets and condensed around the larger colder outer planets (the gas

---

[53] Today gases are trapped below the stratosphere where the temperature drops to below -50°C (220°K).

giants). Nitrogen (N_2) and carbon monoxide (CO) were greatly reduces especially from Mercury and Venus. But the earth was large enough (gravity) and cool enough to retain nitrogen, carbon monoxide, water, ammonia, carbon dioxide, and formaldehyde.

Abiogenesis

The earth cooled and abiotic chemical reactions began to occur. You may be familiar with the Miller-Urey Experiments (1952) in which a mixture of gases (water, methane, ammonia and hydrogen) were subjected to high-energy electrical discharges (simulating lightening) and a number of alpha amino acids (building blocks of proteins) were isolated. I disagree somewhat with that composition being a major contributor to early atmospheric chemistry, primarily because I doubt much free hydrogen existed in the early atmosphere. I favor chemical processes among water, formaldehyde (and the imine of formaldehyde), carbon dioxide and ammonia in water droplets facilitated by solar radiation. Couplings among these gases activated by UV light can account for carbohydrates (sugars):

$$H_2C=O + h\nu \rightarrow H_2C\bullet\text{--}O\bullet + H_2C=O \rightarrow H(O=C)CH_2OH \rightarrow \text{etc.}$$

Of course, 5- and 6-carbon sugars are favored because of the formation of stable hemi-acetal ring systems (C-C-C and C-O-C bond angles ~109°), which terminate the carbohydrate chain growth.

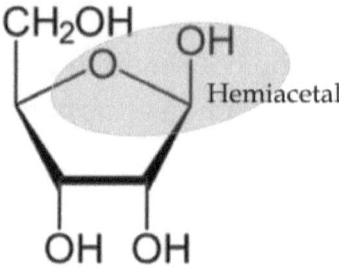

Beta-D-Ribofuranose

Author: NEUROtiker; Source Wikimedia Commons

Similarly, alpha amino acids could arise by a similar pathway:

$$H_2C=NH + h\nu \rightarrow H_2C\bullet\text{--}N\bullet H + CO_2 \rightarrow HO_2C\text{-}CH=NH \rightarrow etc.$$

Cyanate can also be formed from carbon dioxide and ammonia by way of ammonium carbamate. Pyrimidines and purines are likely formed from some of these intermediates and likely "HN=C=O" cyanate, which is known to spontaneously produce ring compounds (cyanuric acid):

Author: Yikrazuul; Source: Wikimedia Commons

Different compounds might be favored in different temperature ranges on the surface of the cooling earth (e.g., higher temperatures might favor purines and pyrimidines over amino acids). Clearly, these compounds would be concentrated in small pools of water left behind when the hot earth reevaporated the water that brought them down.

Perhaps Darwin (1871) was right about this![54]

> *"It is often said that all the conditions for the first production of a living being are now present, which could ever have been present. But if (and oh what a big if) we could conceive in some warm little pond with all sort of ammonia and phosphoric salts, – light, heat, electricity present, that a protein compound was chemically formed, ready to undergo still more complex changes, at the present such matter would be instantly devoured, or absorbed, which would not have been the case before living creatures were formed [...]"*

Of course, all of these amazing carbon-based building blocks would have been wasted had there not been some source of energy to move them into complex products. That energy was likely provided by pyrophosphates produced from phosphorus oxides (e.g., $P_3O_9H_3$) during the molten phase of to the earth's crust. I can imagine pyrophosphates or phosphosilicates (e.g., Bioglass) being released from predominately silicate rock to react with hydroxyl groups on sugars and amino acids.

---

[54] Quoted by Juli Peretó,1 Jeffrey L. Bada,2 and Antonio Lazcano. Charles Darwin and the Origin of Life. *Orig Life Evol Biosph.* 2009 Oct; 39(5): 395–406. Published online 2009 Jul 25.

Evolution

I do not see any mechanisms for the formation of hydrocarbons from likely constituents of the original atmosphere. But I believe that cell walls can be produced from proteins and carbohydrates. Inside the cell, a controlled environment allowed collected molecules to remain concentrated.

Life began in whichever one of these factories could take raw materials found in its environment and enlarge itself. Reproduction followed from fission of cells. Competition was immediate. The cells that could assemble themselves fastest expanded, consumed the natural resources (nutrients), and underwent fission to produce a new generation. Eventually, the preferred resources were consumed, but some cells had a slight advantage and could use a still-abundant less-desirable resource. Metabolism chains and complex mechanisms evolved. Eventually, a molecule that could act as a templet (enzyme) evolved and eventually ways of storing information about how to make the enzymes arose.

The first genome, apparently formed from ribose-phosphate copolymers (which may have originally been part of the cell membrane) joined with a series of pyrimidine and purine bases, were formed. This phase of history is called "RNA world" and this gave way to the use of DNA for information storage. RNA was retained in many important roles, especially formation of proteins.

There are those that contend that it is statistically impossible for life to have evolved without the intervention of an intelligent Creator. This gets into an area beyond science. But I will give my personal opinion so that you will know where I am coming from (and you may freely disagree). Science is very good at answering questions like what, when and how. But we will never answer "why" …why does the universe exist? And therein is where I believe that science and religion part ways. In my view, the fact that anything exists, suggest some truly Infinite Being. And that being, is both the matter and energy of the universe and the laws that make it work. Indeed, in my opinion, it is blasphemy to suggest that the Creator must come in and dabble in the lives for mere humans or in any way for any instant *suspend natural laws*. As I scientist, I am "bearing witness" to the grandeur and perfection of the Creation. I learn is secretes through experimentation and observation and I share that information with my fellows as an act of adoration of the Creator. As a "priest" and "prophet" to fellow mortals, the only and worst sin I could ever perpetrate is to knowingly falsify and mislead them with regard to how the universe works. Indeed, I should caveat *every* disclosure as a *hypothesis*, not ultimate truth. Now that I got that off my chest, what about the idea of "intelligent design."

While it is very true that humans have no statistical chance of evolving an enzyme from scratch, we are the product of a very long line of evolution. We carry the same basic biochemistry developed in bacteria. And we evolve on a higher level (and our evolution is more a matter of physical form than chemical mechanism).

But the original evolution driven by natural selection is not limited by the statistics that humans face. I call it "evolutionary power." Imagine all the water of the planet (10^{20} L) and every liter containing a billion (10^9) evolving cells and every cell reproducing 3 times a day (3) or 10^3 time a year. And imagine this going on for a billion years (10^9). I make that to be something like 10^{41} random events. But wait, there is sex, when something works, it gets shared. So, systems that worked, got shared and were combined randomly with other successful events. With that sort of evolutional power, our DNA with a mere 10^9 base pairs, is *not implausible*.

In contrast, we have not and cannot evolve a major protein from amino acids. And I can prove it. Our "evolutionary power" is a pathetic 10^9 individuals (even today) and over history the effective average number humans is more like 10^3. Excellent logic gives our species a history of less than 10^7 years from the chimpanzees or 10^8 years from the first mammals. But we reproduce only about every 10^1 years. So, we are looking at no more than about 10^7 generations of mammals with an average population of about 10^3, which gives us a mere 10^{10} random events. And most of the mutations of DNA that affect proteins have proven to be deleterious or even lethal. For example, the mutations that cause sickle cell disease are clearly deleterious to the productivity of adults (although they may allow children to survive malaria) and the mutation of the *GLO* gene about 61 million years ago[55] is potentially lethal. Without this gene, mammals cannot make vitamin C and will soon die without a source of ascorbic acid. In spite of the lethality of this mutation, over 61 million years, we have not repaired the *GLO* gene nor found a way around it. Thus, is it safe to say that mammals do not have enough evolutionary power to evolve protein-coding genes and apparently the Creator did not see fit to indulge in "intelligent design" when our species was founded about 10 million years ago.

Evolution of Photosynthesis

It appears that the first life on the planet was dependent on sugar-esters of pyrophosphates to provide the necessary energy to drive chemical reactions. Evidence for this is found in the various phosphate esters involved in the most fundamental metabolic processes (e.g., the Calvin cycle for converting CO_2 to carbohydrates). The original sources of active phosphate must have been inorganic phosphorus oxides (e.g. P_4O_6 to P_4O_{10}), which were volatile when the earth was still hot, or pyrophosphates, which slowly eroded from encapsulating silicate rocks after the earth cooled enough to condense water. Today, phosphorus is found almost exclusively as phosphate (PO_4^{3-}) which is the end product of energy releasing reactions:

---

[55] Lachapelle MY, Drouin G. Inactivation dates of the human and guinea pig vitamin C genes. *Genetica*. 2011;139:199–207.

$$P_4O_{10} + 6\ H_2O + \rightarrow 4\ H_3PO_4 \text{ (aq)}$$

$$\Delta G = [4\ (-1124)] - [1(-2698) + 6(-237)] =$$

$$-4496 + 4120 = -376 \text{ kJ as written}$$

But the general absence of phosphorus in any other form than phosphate (except in the biomass) suggests *that had photosynthesis not evolved before the pyrophosphate supply was consumed, life likely would have ended.* With pyrophosphate becoming hard to find, some bacteria evolved the ability to capture solar radiation to facilitate other types of chemical reactions.

Many books have been written about photosynthesis and scientific papers and recent reviews are available.[56] The overall process is very complex involving two light absorbing complexes and many intermediate steps. Here we will not go into the details of the mechanism.

Fortunately, some primitive bacteria acquired the ability to utilize sunlight for energy. The well-known molecule chlorophyll was apparently the key to capturing light.

---

[56] Cardona T. 2019. Thinking twice about the evolution of photosynthesis. *Open Biol.* 9(3):180246.

Shen JR. 2015. The Structure of Photosystem II and the Mechanism of Water Oxidation in Photosynthesis. *Annu Rev Plant Biol.* 66:23-48.

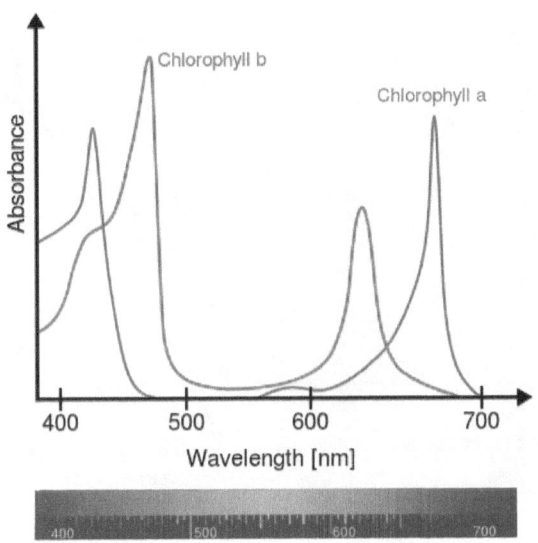

Chlorophyll Author: Yikrazuul; Source: Wikimedia Commons

Author Aushulz, derivative work: M0tty, Source: Wikimedia Commons

It is interesting that the absorption of light by chlorophyll does not match up with the maximum of the solar spectrum reaching the earth (see section 1.4 above). Instead of basing photosynthesis on a molecule that absorbs at, e.g., 550 nm, chlorophyll absorbs at about 475 nm. Why? Recall that photosynthesis evolved in aquatic bacteria. And if we look at the screening characteristics of water, they point to chlorophyll b having been the perfect molecule for optimizing the capture of solar energy in water:

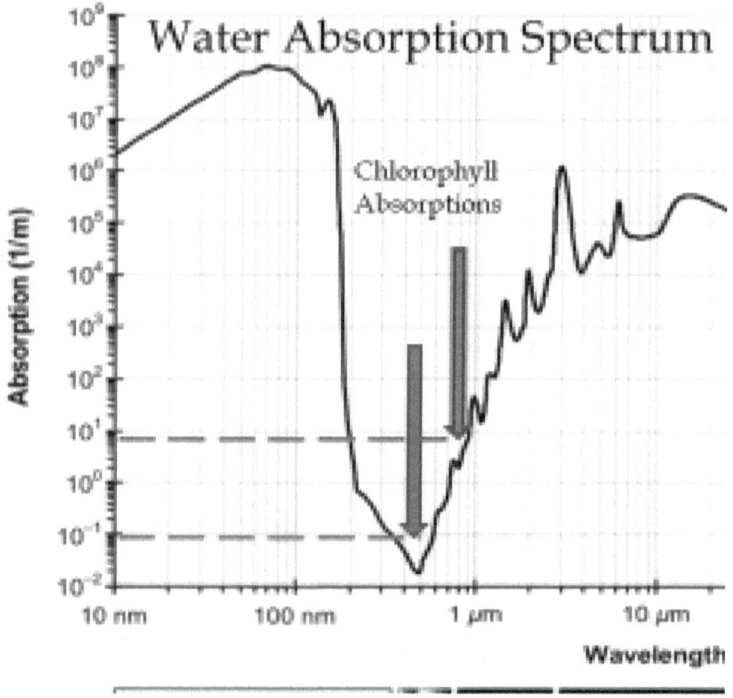

Author Kebes; Source Wikimedia Commons

The absorption spectrum of water is dramatic in that it absorbs strongly over the solar spectrum, except between 200 nm 800 nm with a maximum transmission around 400-500 nm. As you might expect, that is exactly where the main absorption maximum of chlorophyll (a and b) falls.

The absorption of sunlight by chlorophyll appears to have coincided with the evolution of "photosynthetic system I" (Ph I), which incorporated a process for converting low energy ADP into ATP (recovering the phosphate, previously expended):

Author: NEUROtiker; Wikimedia Commons

This system included the use of H_2S (abundant in the ancient ocean) as a source of hydrogen to reduce CO_2 (abundant in the ancient atmosphere and ocean) to carbohydrate "CH_2O" with deposition of solid sulfur (as in modern purple bacteria).

Once Ph I was established, evolution took advantage of the fact that these molecules also absorb light between 600 and 700 nm to develop a secondary mechanism of harvesting solar energy now known as "photosynthetic system II" in which NADH and NADPH emerged as the hydrogen source for reducing CO_2. NAD and NADP were reestablished as NADH and NADPH by stripping hydrogen from water with release of molecular oxygen (O_2). This second region of absorption (600-700 nm) may have become important once plants moved onto the land because the solar intensity radiation is falling off and the water absorption is increasing rapidly in this region resulting in several orders-of-magnitude less energy available under water than on dry land.

Subsequently, land plants have evolved pigments that absorb in the green wavelengths and transfer energy into the photosynthesis machinery (and the color of the leaves in autumn indicates the expression of these pigments as solar radiation wanes).

For example, anthrocyanin absorbs strongly around 500 nm and reflects red light. Similarly, carotenoids absorb strongly slightly more into the blue (200-500 nm) and reflect yellow orange light.

β-carotene

Source: https://harvardforest.fas.harvard.edu/leaves/pigment

Photosynthesis and the Aerobic Atmosphere

The most recent estimates of the evolution of oxygenic photosynthesis (3.6+/- 0.3 billion years ago) and the so-called "Great Oxidation Event" (2.1+/- 0.3 billion years ago)[57] point to a period of time when oxygen was produced locally but not important globally.

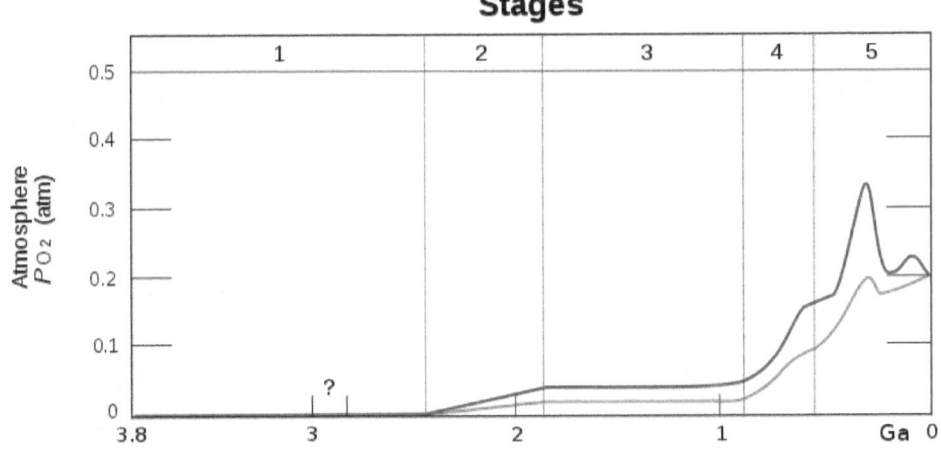

Atmospheric Concentration of Oxygen

(upper limit and lower limit)

Heinrich D. Holland derivative work: Loudubewe;

Source: Wikimedia Commons

---

[57] Garcia-Pichel F, Lombard J, Soule T, Dunaj S, Wu SH, Wojciechowski MF. 2019. Timing the Evolutionary Advent of Cyanobacteria and the Later Great Oxidation Event Using Gene Phylogenies of a Sunscreen. *MBio.* 10(3). pii: e00561-19.

This appears to reflect the fact that the ancient oceans contained large amounts of Fe(II). It took over a billion years for oxygen (O_2) generated in the sea by cyanobacterial (oxyphotobacterial) oxygenation to precipitate most of the iron oxide (Fe_2O_3) as the familiar red clay mineral formations.

After this delay, the oxygen produced in the oceans appear to have established a quasi-equilibrium with oxygen in the atmosphere (0.02 to 0.04 atm) from 1.9 to 0.9 billion years ago during which time oxidation of Fe(II) on the land suppressed build-up of oxygen in the atmosphere. Consumption of that iron (Fe(II)) was complete and the atmospheric oxygen content rose to about (0.1 atm) about 500 million years ago.

Land plants then appeared (pre-ferns and ferns) 0.430 billion years ago and invaded the humid valleys and estuaries of ancient rivers (gymnosperms, 390 Mya). The *carboniferous period* extended from 358.9 million years ago (Mya) to about 298.9 Mya. During this time, the plant material simply piled up, layer upon layer. Simultaneously, CO_2 decreased in the atmosphere, but the solubility of CO_2 in water/the oceans tended to buffer the atmospheric levels of CO_2.[58]

The percentages on the graphic below indicate the sea-level concentration. Keep those data in mind as we go forward. At 5% concentration, there is not enough oxygen for complex animals to survive and plant materials will not burn. There was

---

[58] The solubility of CO_2 in water is about 1.5 g/kg while the solubility of oxygen O_2 in water is only about 0.03 g/kg.

no oxygen to decay or degrade the ferns and ancient trees and fire was not likely to spread in wet domains with the oxygen levels under 15%. (The upper flammability limits of most hydrocarbon gases require about at least 15% oxygen for a flame.)

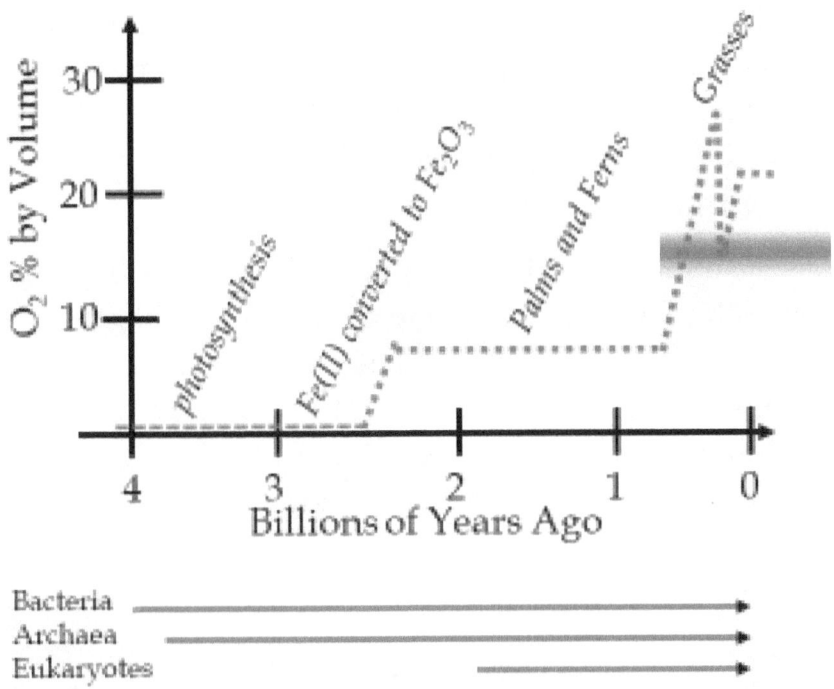

Seed plants and grasses (angiosperms) evolved about 100 million years ago. They were able to widely invade the up-land. Thus, the vast new sources of photosynthesis pushed the oxygen levels if the atmosphere to over 30% (0.30) about 750-million years ago.

This is above the lower flammability limit for oxygen indicated on the graph as a horizontal bar.

When this happened, one bolt of lightning must have started a prairie fire of continental proportions. It was hot and it spread to the edge of the oceans where thick beds of palm fonds were stacked up. Under heat and pressure the organic materials (protected from oxygen) were pyrolyzed. Liquid hydrocarbons were partially decomposed and squeezed out of the organic matter. Naturally the droplets of organic liquids adhered to grains of sand and silt and were washed own stream into the oceans where they settled out in shallow seas.

The fires raged for several hundred million years. The oxygen levels fell until continental fires were no longer possible and slowly an equilibrium has been established between periods of fire (declining oxygen levels and increasing CO_2 levels) and periods of growth (increasing oxygen levels and decreasing CO_2 levels). Of course, local climate and humidity tend to randomize and localize these cycles.

Chloroplast, Mitochondria and Eukaryotes

The interesting (if not ironic) factor here is that that photosynthesis produced oxygen as a byproduct and oxygen was arguably the first global toxic material. Oxygen is very reactive and is lethal to many (anaerobic) bacteria. There appears to have been an evolution of highly specialized cells that conducted photosynthesis assimilated hydrocarbon compounds and released oxygen.

These organisms seem to introduce "coenzyme A" (CoA) and acetyl-CoA, which along with the reducing power of NADPH forms systems capable of (i) adding an acetyl group to acetyl CoA, (ii) reducing carbonyl compounds to the corresponding secondary alcohol, (iii) eliminating the H/OH to make a double bond, (iv) and reducing the double bone to an ethyl group:

CoA-CO-CH$_3$ → (i) CoA-COCH$_2$-**COCH$_3$** →

(ii) CoA-COCH$_2$-**CHOH-CH$_3$** → (iii) CoA-COCH$_2$**CH=CH$_2$** →

(iv) CoA-COCH$_2$**CH$_2$CH$_3$** → (v) CoA-COCH$_2$**COCH$_2$CH$_2$CH$_3$**

And subsequently, moving the entire "alkyl- or alkenyl-acetyl" group to another CoA-acetyl and going again. This was the beginning of synthesis of fatty acids, alkenone[59] and other hydrocarbon chains with single and double bonds (through similar processes: squalene, carotene, isoprene, steroids, terpenes).

The evolutionary pressure caused by the constant exposure to oxygen, which caused mutations in the DNA of these cells, resulted in very degenerate cells with a minimum of functionality. Meanwhile, the chemical utility of oxygen as a nutrient in an environment (composed of reduced metals and

---

[59] A bioproduct of marine algae (*coccolithophore Emiliania huxleyi*) which began at least 0.120 billion years ago. These are found in ocean sediments and uses to estimate historical Sea Surface Temperature.

hydrocarbon compounds) kicked off a branch of evolution to organisms that specialized in utilizing molecular oxygen. The handling of molecular oxygen also degraded these highly specialized bacteria. In the end, the specialized bacteria became what we recognize as *chloroplast* and *mitochondria*.

These rudimentary cells were captured by various lines of non-specialized bacterial cells and produced a branch of the family tree known as the Archaea. Lynn Margulis (born Lynn Petra Alexander; 1938–2011) was the first to advocated for a *symbiotic contribution to evolution*. The Bacteria continued to diversify into many species; and within the chaos of the Achaea, a dramatic improvement was achieved: Linear chromosomes were incorporated in a nuclear membrane where they could be protected from the oxygen-based metabolism going on in what became the *cytoplasm*. These organisms are known as the Eukaryotes (1.6 and 2.1 billion years ago). These cells evolved mechanisms for differentiation (cellular metamorphose) and gave rise to multicellular organisms with specialized cell types flourished: Plants and Animals.

Coal and Petroleum

Coal clearly is a residue of massive layered plant materials. We can still find imprints of palm fonds and tree stumps in coal beds. One that dates back 307 million years ago has been found 250 to 800 feet underground in a southern Illinois coal mine. It is well understood that chlorophyll (which gives the green color to leaves) is one of the key molecules involved in collecting radiant energy for photosynthesis (see structure above). You will notice that it has a long ester sidechain connected to the porphyrin ring system by a carboxyl group. This sidechain is hydrolyzed and degraded anaerobically to phytane:

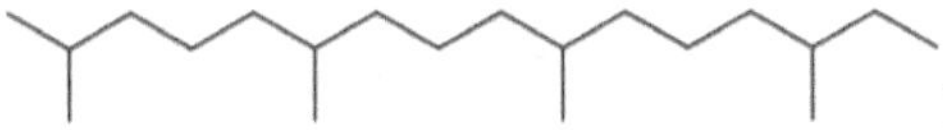

Phytane

Author: Edgar181; Source: Wikimedia Commons

Under aerobic conditions it is degraded to pristane (which has one less carbon):

Pristane

Author: Edgar181; Source: Wikimedia Commons

While the oxidation may occur in different environments, there is no doubting that these hydrocarbons originated from green plants.

As expected from the hypothesis outlined above, we do find oil deposits in sand near the mouths of large rivers that drain areas rich in coal (e.g., the Caribbean Ocean receiving sand from the deposits in the Missouri and Ohio River valleys; the North Sea receiving sand from the Humber and Rein River valleys).

The Human Species

What is a species? This is a question that has complicated the understanding of biology and evolution until very recently. Originally, species were defined by morphology (appearance). I do not look anything like a bird or a turtle so I must be in a different species. Well into the 1800s, scientists were arguing about whether Africans and Europeans are in the same species. These types of arguments have led to mass confusion and a condition we call *racism*. In the racist paradigm, "lower forms" (defined by less technological advancement) lead to "higher forms" (defined by higher technological achievement). BUT THAT IS NOT THE WAY IT WORKS. Humans have evolved from a common ancestor to adapt to our environments. So substantial differences in the morphology (and technological achievement) of groups of people can exist *within a single species*.

It was not until the understanding that the chromosomes control inheritance in the 1940s that Ernst Mayr (1904–2005) argued for the "biological species," namely all the mutually fertile individuals belong to a single species regardless of morphological differences. What matters is that they have systems of chromosomes that allow reproduction among all individuals. Thus, since it is now established that "modern humans" (like myself) carry DNA that was evolved in my ancestors the Neanderthals and Denisovans, we are in fact all part of the same species. Of course, that includes all "modern humans" on the planet regardless of morphology, technology or locality. The point is that as long as a group can reproduce, there will exchange of genetic information and the offspring will continue the evolutionary process.

I have published a hypothesis that inversions of centromeres (rare but real mutations) are the key to speciation. [60] Here I disagree with Darwin who believed that *evolution alone* caused speciation. This was the error that allowed racism to drive human relations. In fact, it is speciation that isolates groups genetically and allows massive evolution to occur. Simple physical separation and isolation of groups (e.g., on different continents) is not enough to ensure speciation. Thus,

---

[60] Parris GE. 2011. The Hopeful Monster Finds a Mate and Founds a New Species. *Hypotheses in the Life Sciences*. 1(2):1-6.

Parris GE. 2013. Application of a Hypothesis to Speciation in Hominidae. *Hypotheses in the Life Sciences*. 3(1). I have previously hypothesized that biological species are not the result of gradual changes in the genome or morphology as suggested by Darwin, but rather the result of punctuated major pericentric or paracentric inversions or other rearrangements (e.g., chromosome fusions) that prevent reproductive compatibility with the parent group. Following the rearrangement, a new nascent species can be formed through inbreeding within two generations consistent with the views of Goldschmidt. Applying this hypothesis to speciation in Hominidae (the great apes) suggests that (i) orangutans are close to the common ancestor of Hominidae; (ii) humans are close to the common ancestor of Hominoids, which was adapted for efficient all-terrain locomotion; (iii) gorillas and chimpanzees have passed through more species as they have adapted to a very specialized ecological niche in the tropical forest, and (iv) speciation events in *Homo* facilitated evolution of the human brain.

European/Neanderthals, Africans and Asians/Devonians are the same species adapting to different circumstances.[61]

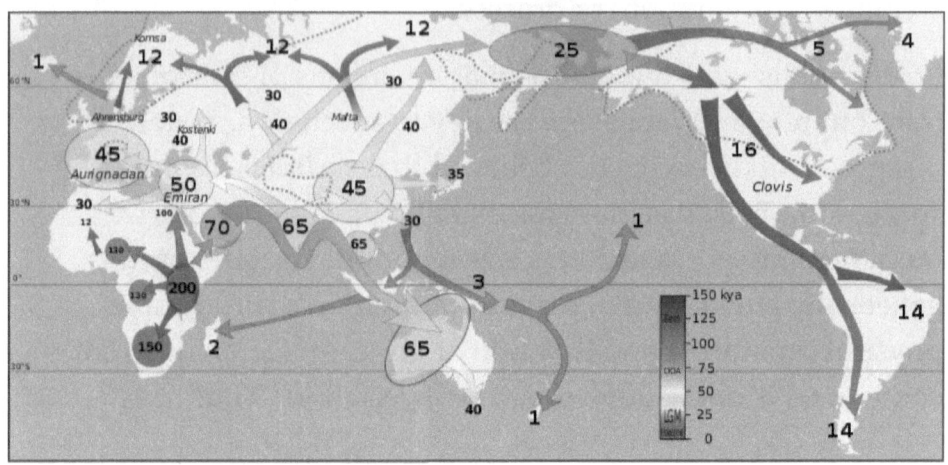

Human Migrations for the last 150,000 years.

Author: Dbachmann; Source: Wikimedia Commons

The main point is that technological progress cannot be equated with evolutionary progress. We will soon see that some environments encourage and facilitate technological advancement and other situations virtually preclude it.

---

[61] It follows from my hypothesis that speciation events have occurred about ever 1 million years in humans, which is about the time modern humans displaced *Homo Erectus* (before the separation of Neanderthals and Devonians.

Geology and Geography

The oxides of aluminum, silicon and phosphorous form polymers that are relatively high melting and less dense than molten iron. They, thus, floated to the surface of the earth and as the earth cooled; and formed crystalline rock of various sorts with inclusions of many different metals and nonmetals. Between the crystalline surface and liquid core there is a viscous (molten glass) mantel. The atmosphere cooled and condensed into oceans and by about 500 million years ago the crystalline rocks only protruded above the oceans in a large continent called Pangea.

This was the beginning of the Paleozoic Era (541-252 million years ago.) This corresponds with the invasion of the land with grasses and the spike in oxygen concentrations in the atmosphere and in the oceans. The result was an explosion in evolution that brought forth a wide class of complex multicellular organisms. This is called the "Cambrian explosion" and was fueled by extensive mutations caused by high oxygen content and the availability of oxygen to generate energy for mobility as well as metabolism. The Ordovician period saw the first vertebrates (internal skeletons) and the Silurian period saw the introduction of jaws and claws to catch and eat pray. This opportunity for actively capturing food naturally favored vision and mobility. Meanwhile the seed-bearing plants ushered in the Devonian and Carboniferous period, pushing the oxygen levels ever higher.

Continental fires broke out and oxygen levels plummeted in the atmosphere and the ocean. This may explain the extinction of highly mobile (oxygen requiring) but inefficient aquatic species of Placodermi and trilobites. Only fish and arthropods with efficient gills could survive at lower oxygen levels. Oxygen fed hot fires and the piles of biomass that had accumulated in previously swampy areas were pyrolyzed and CO_2 returned to the atmosphere.

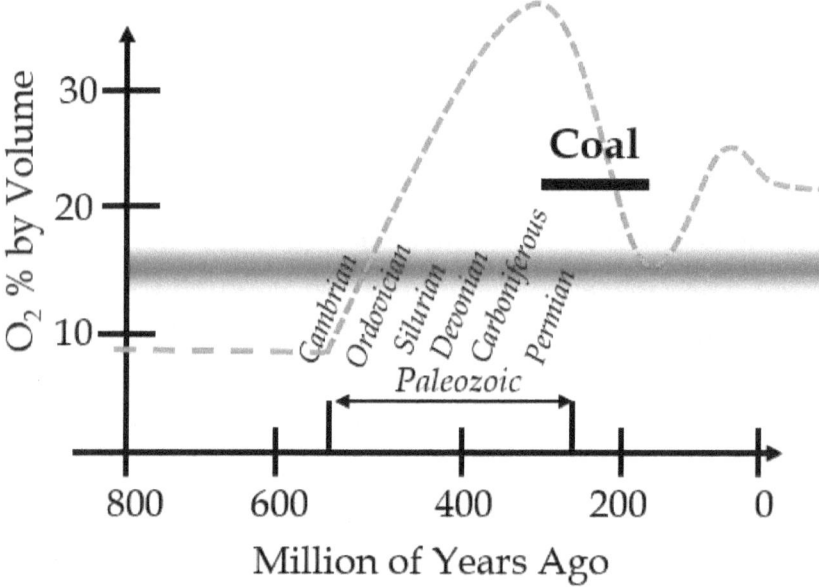

Meanwhile, the dynamic balancing of the spinning earth dictated that the land masses must move. Pangea fragmented and the landmasses drifted on partially submerged plates. Collisions of these plates results in kinetic energy being turned into frictional heat. These melting zones produce an abundance of tectonic (earthquake) activity and volcanoes. Subsurface intrusions of liquid silicate polymers from the lower (hotter) mantel also occur in areas where the overlying crystalline rock is relatively thin. These hot spots may remain active for many millions of years and periodically burst through the crystalline rock as the continental plate moves above. The best-known examples are the chain of atolls and islands (from French Frigate Sholes to the Hawaiian Islands across the pacific; and the series of volcanoes associated with Yellowstone National Park):

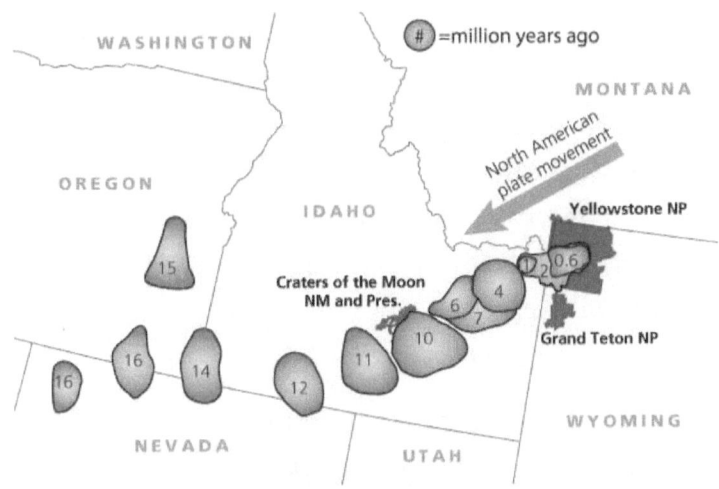

Source: National Park Service US Government

The magma created at a hot spot or along a continental merge, represents many tons of molten rock. Within that rock is a significant amount of water, sulfur and a variety of trace elements (metals). The water (under high pressure and high temperature) is superfluid[62] dissolves the trace metal sulfides and gathers them near the top of the magma intrusion. If a crack develops leading to the (much cooler) surface, any molten metals (such as copper, silver and gold) that do not readily from oxides or sulfides and the hydrothermal waters will certainly move up the crack and cool along the way.

The least soluble minerals start precipitating. This usually included quartz (silicate) with traces of iron. At high temperature these deposits are colorless, but at lower temperature (less than 500°C), the iron atoms can become trapped in the lattice in spaces normally occupied by silicon atoms. This is a very unusual coordination geometry for iron as Fe(II or III) and they are unstable in this environment. Thus, the iron can easily be oxidized to ferrate (Fe(VI) surrounded by a tetrahedron of oxygen atoms), which gives the quartz a purple color (amethyst). Heating the crystals (300-400°C) or exposure to ultraviolet light allows the iron to move to a preferred site in the lattice and the color is lost.

Thus, gold (and less commonly silver and copper) are deposited in veins of quartz. Erosion of these rocks produces malleable pieces of metals that accumulate as nuggets.

Other metals (such as lead) form stable sulfides that are deposited at relatively high temperatures including galena (PbS),

---

[62] *Superfluid* means that the molecules are in contact as in liquids, but they have more than enough kinetic energy to separate (as a gas) under normal pressures. In general, superfluidity means that the material is a liquid with virtually no viscosity.

chalcopyrite ($CuFeS_2$), pyrite (FeS_2), molybdenite (MoS_2), bismuthinite (Bi_2S_3), and arsenopyrite (FeAsS). Stable oxides of metals including tin, uranium, antimony, iron (magnetite Fe_3O_4) and mixed minerals of iron, manganese and tungsten also precipitate at relatively high temperatures.

At lower temperature (200-300°C) carbonate minerals[63] (calcite, dolomite) are precipitated and at very low temperature (50-200°C, nearest the surface the more soluble minerals are deposited in sulfides of arsenic, antimony, copper, silver, and mercury. The halides, nitrates, borates, and other common soluble salts are typically discharged in surface water and runoff to the oceans.

Now that we have an oxygen-containing atmosphere, the surface minerals are subject to oxidation: sulfides are frequently oxidized to sulfates, arsenic ends up as soluble As(V) oxides. But antimony, bismuth, lead and tin sulfides end up as insoluble oxides. Because of the great density of these oxides, they tend to accumulate as pellets segregated by washing processes (e.g., tides) from silica sand.

---

[63] Carbonates are thermally unstable and hydrothermal deposits are sparse. Carbonate minerals of metals like copper are more likely to be formed from atmospheric carbon dioxide acting on copper oxide or hydroxide, which arises from weathering of copper sulfide hydrothermal deposits.

Erosion and Sedimentation

The crystalline rocks (primarily silicate) are broken and worn down by freeze-thaw cycles, flowing water and wind. The small particles are moved by suspension in moving water and air and are deposited layer upon layer. This idea of *superposition* (oldest layers at the bottom) has become a keystone in geology and archeology. One point is worth mentioning here. When the continental plates move, the layers can be folded to the point that in some places the layers are almost vertical. Stone has great compressive strength, but breaks fairly easily under tension, so each layer tends to rupture on the edge that is on the *outside of the curve*.

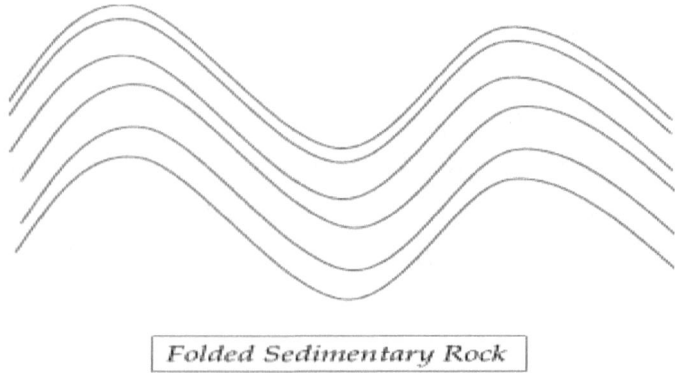

Folded Sedimentary Rock

Thus, the tops of anticlinal mountains are fractured and erode to become valleys, while the synclinal valleys are left standing as plateaus rising above the sedimentary valley.

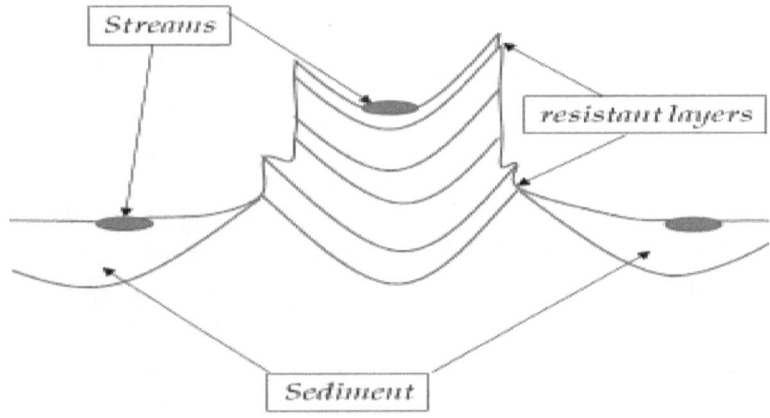

Cloudland Canyon, North West Georgia

Photo by Eric Champlin (2013) for Atlanta Trails

Resistant layers are frequently sandstone or rich in iron oxide while the rapidly eroded layers are limestone. Caves may develop in the limestone as water trickles down through cracks. It is not unusual for coal layers to also be presents in these hills (low flat-top mountains, e.g., tepui). The streams that collect on the top of the mountains may create magnificent waterfalls if the rock beneath them is impenetrable.

Angle Falls from the
Auyan Tapui
Venezuela

Neolithic Technology

Soon after the introduction of agriculture (12,000 years ago), humans were able to establish permanent villages. And soon there was a need to store water, grain and personal items. The first step towards a science of materials came with the skill of pottery making. Sundried pots and bricks gave way to the idea of firing them in a camp fire.

Wood (ordinary dry) begins to burn at about 240°C (460°F). A wood camp fire typically burns much hotter near 500°C (930°F) because the rising air creates a draft. It turns out that in a combustion system the rate of heat generated is limited to the rate at which oxygen can come into contact with the fuel. Thus, turbulent flow of air is desirable (so that oxygen does not have to diffuse through laminar layers). However, the rate of heat loss includes the heat lost by the exhaust gases (mainly nitrogen and carbon dioxide) and the radiant heat lost directly from the fire. So, there is a limit to the utility of faster gas flow. Larger fires (bonfires) have less surface area per unit volume of combustion and get hotter (up to 1,000°C). Pure silica melts about 1,715°C, but silicate-based minerals (including sodium, calcium, magnesium, potassium ions, etc.) generally melt between 600°C and 1200°C. Thus, some crude silicate glasses can be formed in bonfires.

Ancient peoples could not do much about the heat lost by the exhaust gases, but by trial and error, they discovered that a clay chimney both focused the natural draft and minimized the radiant heat loss. Soon the pottery industry included kiln-fired pots:

Kiln Color	Approximate Temperature
Orange-Yellow	950°C (1750°F)
Yellow	1200°C (2300°F)
Bright Yellow	1330°C (2500°F)

In these ranges, clay pots come crude glass pots as the individual grains for clay fuse together.

Addition of the minerals of copper, lead and other metals resulted to a surface coating that had distinctive colors.

5.2 The Beginnings of Modern Materials

Native Metals

Gold is typically found in nuggets of the pure metal. Gold was undoubtedly one of the first metals recognized by humans and it was easily worked into jewelry by moderate heat or forging. Native (elemental) copper and silver were known. In particular native copper is found in upper Michigan on the peninsula between the great lakes. Ironically, the pieces are rather large, which makes it difficult to try to melt in a camp fire. There is evidence that copper nuggets were mined and distributed in the region (probably traded throughout North America), but the metallurgy of copper does not appear in the Americas until the arrival of Europeans. When the Europeans arrived, the Aztecs were armed with wooden swords with obsidian (volcanic glass) imbedded in the edges.

The following table of data will help explain how technology developed from here:

Properties of Ancient Metals and Alloys			
Metal	Melting Point (°C)	Typical Hardness (MOHS)	Decomposition of Oxide (°C) To metal
Mercury (Hg)	-39	Liquid	500
Sodium (Na)	98	0.4	>1100
Tin (Sn)	232	1.5	>1900
Lead (Pb)	326	1.5	>1477
Zink (Zn)	419	2.5	1974
Antimony (Sb)	632	3.0	>1425
Aluminum (Al)	660	2.9	>2072
Brass (Cu/Zn)	930	4	
Bronze (Cu/Sn)	950	4	
Silver (Ag)	962	3.0	~300
Copper	1084	2.5	>2000
Gold	1064	2.5	~300
Iron (Fe)	1535	4.5	>1539
Wrought Iron	1500	4	
Steel (Fe/C/etc.)	1370	Up to 9	
Obsidian (volcanic glass)	1710	6	>1710

Metallurgy in Early Mediterranean Civilization

Western history is Eurocentric. I apologize in advance for giving little attention to East Asian and sub-Saharan African advances and contributions to modern materials science. There appear to have been some important independent discoveries in the very early history of our species starting about 10,000 years ago. By focusing on the Eurocentric advances, I am following only one pathway and I do not mean to infer "priority" in the sense that is usually considered in inventions. Nonetheless, I think it is fair to point out that we can probably never prove whether what we recognize as European was actually introduced from other cultures; and, more importantly, the European history of technology could have reached today's status without other contributions. So, here we are.

Based on what I can find regarding geology, metal ore deposits and historical records, the eastern Mediterranean was the focus of a civilization beginning with cultivation of crops and permanent settlements about 10,000 years ago. The Middle East is a crossroads of peoples and a collision point of continental plates. The result is a wealth of prehistoric vulcanism and hydro-geothermal activities that deposited trace metals in economically useful quantities in a variety of places including around Lake Van and Lake Urmia in ancient Armenia, Wadi Faynan (south of the Dead Sea) and the entire island of Cyprus.

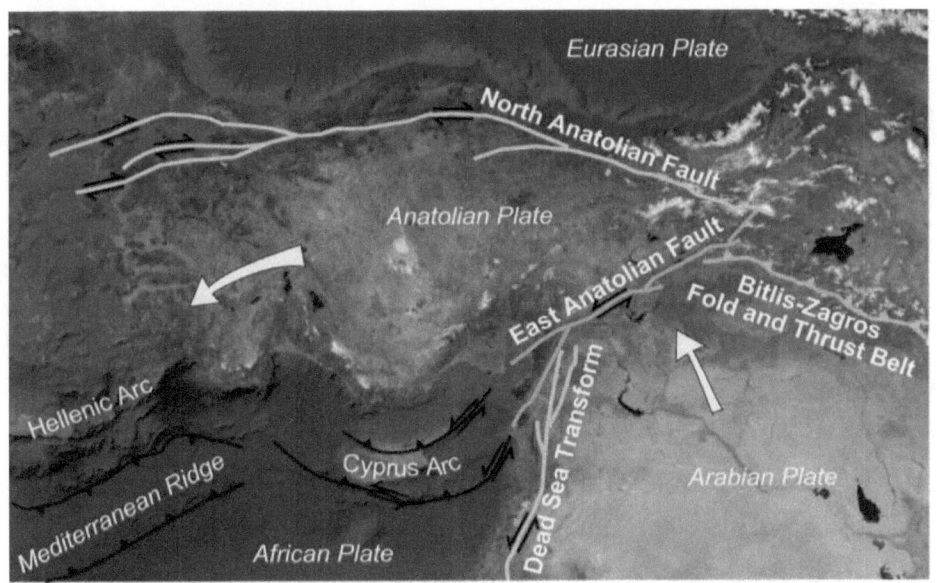

Author: Mikenorton; Source: Wikimedia Commons

Cyprus is a not a typical situation. As the African plate was forced under the European plate an under-sea (anerobic) mineral deposit was scrapped off and brought to the surface in the last 5 million years. The sulfide minerals have, therefore, weathered relatively little (i.e., the minerals have not been exposed to erosion and oxygen as much as other places) and as a result many of the minerals on Cyprus are still sulfides of iron and copper:

$$(\text{chalcocite}) \; Cu_2S + O_2/H_2O \rightarrow Cu_2O$$

$$(\text{covellite}) \; CuS + O_2/H_2O \rightarrow CuO$$

$$(\text{cuprite}) \; Cu_2O + O_2 \rightarrow CuO$$

$$CuO + CO_2 \rightarrow CuCO_3/Cu(OH)_2 \; (\text{Azurite and Malachite})$$

Copper sulfides were likely the first sources of copper metal because the colorful minerals can be directly smelted to copper in the orange-yellow embers of a bonfire:

$$CuS/Cu_2S + O_2 \rightarrow Cu + SO_2$$

Thus, copper became one of the first metals to be produced from a mineral ore and it was soon being melted and recast in new shapes. But it had a major limitation because it could not be used for tools. Like gold and silver, it was very malleable and it tarnished when exposed to air. To make an edged tool (ax, shovel) or weapon (spear, sword) or a fastener (nail or rivet) a metal needs to have a hardness of at least 4 (MOHS). Thus, Cyprus may have been the best source of copper in the earliest days, but it was probably not the pace were the copper ore was first turned into useful implements.

The science of metallurgy appears to have had it western origins in ancient Armenia, specifically, the region around Lake Van (now in Turkey) and Lake Urmia (now in Iran). These were relatively isolated mountainous regions and seemed to maintain independence from both the Hittites and Assyrians by providing metals (gold, silver, copper, lead, tin) in trade and in taxes. Whether or not it was first, is immaterial, but 3000 years ago, the technology of reduction of metal oxides with carbon was well understood and widely practiced:

$$CuO + C \rightarrow Cu + CO_2$$

$$PbO_2 + C \rightarrow Pb + CO_2$$

$$SnO_2 + C \rightarrow Sn + CO_2$$

Once knowledge is acquired, it spreads rapidly if there is a demand for the product. Thus, reduction of metal oxide ores with charcoal was practiced throughout the region. I suspect that wood (for making charcoal) was a limiting factor on Crete and in the Levant.

Bronze Age

Serendipitous alloys of copper occasionally produced a metal with superior hardness. But with tin and copper being produced in the same region or Armenia, bronze (90% copper, 10% tin) was undoubtedly discovered. Bronze is easy to melt (950°C) and cast; and it will take an edge.

Highly organized societies were established before 3000 years ago (1000 BCE). I will focus on the Egyptians, Hittites and Assyrians. The Egyptians, of course, occupied the Nile River Valley and delta; the Hittites were firmly established in Anatolia (modern Turkey) and the Assyrians occupied the Tigris and Euphrates valleys. These empires developed systems of government/laws, writing and technology needed to build cities and roads and make war. In fact, the Egyptians and Hittites were in extended conflict and fought one of the great battles of history at Kadesh in 1275 BCE. After the battle, one of the world's first written treaties between the Hittites and Egyptians created a buffer zone known as Palestine/the Levant along the eastern shore of the Mediterranean Ocean.

The power of these nations was dependent upon the men, arms and armor that they could mobilize; and the arms and armor depended on bronze. With the enormous resource of copper in Cyprus, the ability of these empires to make bronze was limited only by the availability of tin. Very soon, the call went out on every trading route for tin.

Tin is much less common than copper. It is likely that it came in from all directions *in small quantities* from 3000 BCE until about 1100 BCE.

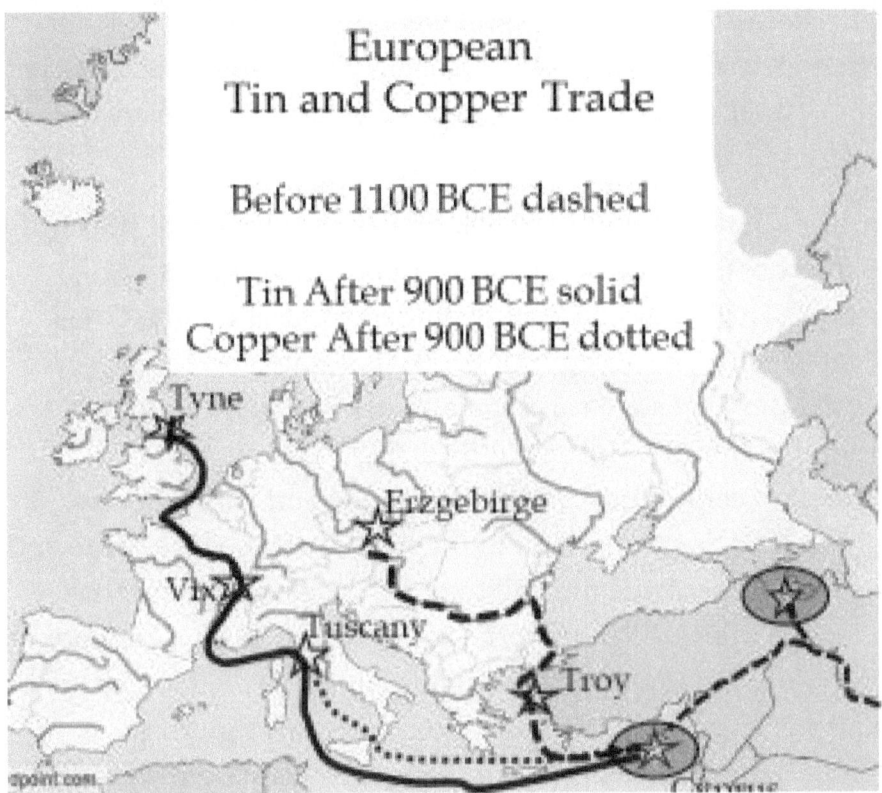

The Mycenean civilization (Knossos) gave birth to the Greek city states, which almost immediately went to war with Troy (a trading post of the Hittites providing access to Europe via the Danube River), probably for trading access into the Black Sea. It is generally agreed that some sort of calamity struck the Middle

East about the time of the Trojan War. To this point, archeological discovers point to a period of about 200 years (1100-900 BCE) during which all the empires including the new Greek states and Troy ground to a halt: Writing was lost, art reverted to simple designs, and commerce seems to have been frozen. Disease, famine, war, and weather have been named as contributors to the disruption. Before 1100 BCE tin was imported into the copper producing regions from central Europe and eastern Asia.

The history of the period 1100-900 BCE is cloudy, but at the end of the darkness, a new empire arose in north-west Italy (Tuscany). The Etruscans likely had historical ties to the Greek City States, Hittites and Egyptians. They established a government of city states and metal-working tradition. They also founded the city of Rome (753 BCE). I believe that they explored into central Europe via the Rhone and Saone Rivers, crossed the continental divide to the head waters of the Seine and ultimately followed the tin trail to north eastern England (the Tyne River Valley). This hypothesis is supported by a common cultural tradition of chariots and "chariots burials" that can be traced from Tuscany, to Vix, and from Vix via Belgium to Yorkshire (Humber River) in England. It is a mere two-day walk from Vix on the headwaters of the Seine to the headwaters of the Saone River. Archeological examination of Vix has revealed a Bronze Age culture (circa 600 BCE) far exceeding any expectations of the native Celtic tribes of Gaul.

Source: Musée du Pays Châtillonnais

This Krater found at Vix, France on the headwaters of the

Seine River is 1.63 m (5'4") Tall

It appears that, the Etruscans organized tin mining and metallurgy in the Tyne Valley and actively exported tin to Tuscany, Cyprus and the Levant. Ships returning from Cyprus would be expected to bring Copper and bronze back to Tuscany (900-600 BCE).

The conventional hypothesis that the Phoenicians imported tin and copper from Cornwall to the Middle East makes no sense whatsoever. First off, no one would import copper to Cyprus or the Levant. When Julius Caesar explored Gaul (55 BCE), his commentaries were written for political impact in Rome; but it is clear that there was not a massive trade in tin or copper from Britain to Rome in 55 BCE. Indeed, the Romans (55 BCE to 410 CE) never even occupied Cornwall and Wales was primarily a source of annoyance to the Romans. In contrast, Rome occupied the Tyne River Valley and built Hadrian's Wall (starting in 122 CE) to protect it from attack by the Scotts.

Caesar had been attempting to depict himself as a discoverer of a new land. Indeed, he may not have known the history. The reason that Rome was so ignorant of Gaul and Britain by 100 BCE is the fact that (i) iron smelting and wrought iron weapons had begun to replace bronze and (ii) when Hannibal Barca (247–181 BCE) had swept from Spain into northern Italy (220-200 BCE), he had effectively cut off Italy from its trading partners in Gaul. The old trade routes were obsolete and the Etruscan/Roman *ex patriates* in Britain and Gaul merely coalesced with the much larger local Gaelic tribes.

The history of copper and bronze provides an interesting example of the effects of economic demand on advancement of trade and technology. We will see this effect throughout history especially the impact of the world wars of the twentieth century (1914-18) and (1939-45).

Mining Technology

It is worth mentioning mining technology at this point because the ability to mine naturally occurring minerals plays a role in the history of metallurgy. The point that is to be made is that *metal production stops when the technology that is available to extract the minerals no longer works or is economically infeasible.* In the ancient world, there were no effective ways to pump water out of mines. In a few unique cases, the water table could be lowered by digging horizontal shafts into cliffs.; but for the most part, mining and metal production stopped at the water table. Of course, as soon as effective pumps become available (e.g., 1700 CE), the mineral explorations resumed.[64]

For example, look briefly at production of tin and copper in Britain. Bronze was initially imported into Britain (before 2400 BCE). Small copper deposits (e.g., Ross Island, Ireland) were found and exploited by 2100 BCE. Small local deposits of copper were found and quickly depleted. Economical copper ore was relatively high grade and was found in veins associated with hydrothermal vents. To extract the ore miners typically had to dig progressively deeper into the earth. Children were often employed for this task because they are fearless and because they are small enough to go through narrow tunnels.   The "Great Orme" mine in Wales was dug about 1800 BCE (3800 years ago) largely by children to a depth of 70 meters (230 feet) with about 8 km (5 miles) of tunnel.

The geology of tin, antimony and lead tends to follow a different pattern from copper. These metals are deposited relatively near

---

[64]That is part of what happened in Cornwall, England. It may have had a mineral industry in 1000 BC, but the minerals were inaccessible from 600 BCE until 1700 CE.

the surface of hydrothermal vents and erosions exposes them to atmospheric oxidation. Thus, the elements collect as dense[65] insoluble oxides that agglomerate and are sorted into alluvial seams as the less-dense silicate sands wash away (like panning for gold). When appropriate ores are found they are typically enriched by washing out the lighter minerals leaving behind the concentrated cassiterite (SnO_2) mineral.

Alluvial tin crude ore

Alluvial tin concentrate

http://www.frmining.com/Product/Ore_Dressing

/Tin_Extraction_Machine.html

---

[65] Density of metal oxides: PbO_2 (9.3 g/mL); Sb_2O_3 (5.2 g/mL), SnO_2 (6.9 g/mL) as compared to silicate sand (2.65 g/mL)

Indigenous bronze production in Britain involved bringing together the copper from Wales with tin from Cornwall and Devon. Some bronze was exported into north-western Europe; but on a global scale, this was a minor activity until the early industrial revolution.

I estimate that around 900 BCE the economies of the Greek, Persian (successors to the Hittites and Assyrians) and Egyptian empires rebounded and trade boomed. The demand for bronze reached new heights and the search for tin expanded throughout Europe. During this period, the Etruscans appear to have found a more viable source of tin in northeastern Britain (Yorkshire, Durham) and the Tyne River Valley. It has only been since the early industrial period that new technology allowed the resurgence of mining in the southwest of Britain.

Charcoal and Turpentine

If you are reducing, metals on a commercial scale, you need a lot of carbon. Of course, wood could be burned, but harvesting trees and transporting wood into population centers is laborious and inefficient. The moisture content of a tree is approximately 50%. Even lumber after partial drying can have 20% of its weight in water. In addition, there are oxygen containing compounds (cellulose) in wood that have less heating potential per unit mass than carbon. Thus, what has typically happened is that at some point (after land is cleared for crops near villages) hauling wood into town for heating or industrial purposes is just not feasible.

This is where the charcoal-maker steps in. To make charcoal, logs are piled up and covered with earth with just enough openings to allow a smoldering fire. The logs are ignited and allowed to burn with careful limitations on the amount of air admitted. Frequently a liquid sap (turpentine) can be drained away as it is cooked out of the logs and ultimately the wood is completely pyrolyzed to carbon (without burning). The charcoal is nearly pure carbon and has high heat content. Turpentine obtained as a byproduct (principally pinenes) can be used as an organic solvent and liquid fuel.

The conversion of wood to charcoal is a much-overlooked process allowing the construction of large cities, which required heating.

Olive Oil and Light

In the ancient world, light at night could be provided by wooden torches soaked in turpentine. In the Middle East, the olive tree was discovered to produce enough oil to be useful for lighting lamps for use in the evening hours. But the oil was very expensive because the oil was in such limited supply. Other natural oils and waxes (whale oil, bee's wax) were also limited in supply.

This was to prove to be an issue into the 19th century of the current era.

5.3 Iron and Steel

Unlike copper and tin, iron is found everywhere. Of course, some of the geothermal ores of copper contain iron and after thousands of years of copper metallurgy, it was known that there were some materials (possibly metal ores) that were separated from copper. And undoubtedly, people attempted to smelt and reduce the red iron ore that was found in many places.

Iron Age

We frequently associate iron age technology with Britain because ultimately this was one place where its technology flourished. But the basic forced air system depicted below was invented in many places and was still being used into the 1800s by African tribes. But it is noteworthy that iron bars were one of the principal items imported into Africa in the 1800s as part of the slave trade. Native smiths could turn these bars into tools on this forge more economically than starting with iron ore themselves.

The problem with iron is that it melts at higher temperature than can be achieved with a simple forced-air furnace. If a mixture of charcoal and iron oxide is combusted, there is reduction of iron oxide to iron, but because the particles of iron do not liquify and run together, the constant stream of air simply re-oxidizes finely-divided iron to iron oxide. Eventually, through trial and error, a system that more-or-less works was developed:

As shown in the diagram (below) the bottom of the furnace is charged with pure charcoal with the expectation of producing mainly carbon monoxide (CO) and, importantly, depleting oxygen from the air.   The carbon monoxide moves upward and takes oxygen from the iron oxide.

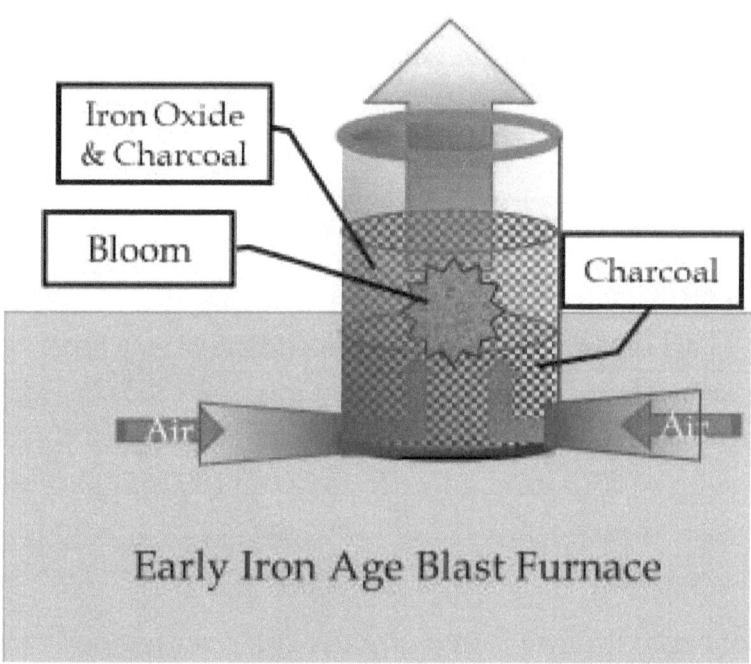

Iron Oxide & Charcoal

Bloom

Charcoal

Air

Air

Early Iron Age Blast Furnace

The tiny particles of iron (Fe) produced likely react with the carbon monoxide to form a volatile iron pentacarbonyl (Fe(CO)$_5$) complex, which is volatile (bp 103°C) but unstable at this temperature. The transient existence of this volatile compound allows the iron atoms to be welded together without ever being a liquid. [66]

$$C \text{ excess} + O_2 \rightarrow CO$$

$$3\ CO + Fe_2O_3 \rightarrow 2\ Fe + 3\ CO_2$$

$$Fe + 5\ CO \leftrightarrows Fe(CO)_5$$

The result is formation of a blob of iron and carbon that can be recovered as a solid from the furnace. This mass is called a "bloom." The hot bloom is beaten with stones or metal hammers (i.e., forging) to allow the malleable iron to collapse while the brittle charcoal/carbon is extruded to the surface. Re-heating the mass burns off the carbon and allows farther compaction of the iron by hammering. The process of forging is continued until a bar of iron with relatively little carbon (mostly alloyed with the iron on a molecular scale) is achieved. The product is "wrought iron." The process could be continued to the point where almost no carbons is present. Indeed, wrought iron can become steel (e.g., Damascus steel).

The term "iron" is applied to iron with over 2% carbon. The term steel is applied to iron with lower amounts of carbon, but the propertied of steel is still very variable: Mild steel (less than 0.35 % C) is easily welded but by the time the carbon contend reaches 0.5 %, the steel is almost un-weldable (very prone to cracking). As the carbon content increases the steel becomes

---

[66] It is relevant that in 1890 Ludwig Mond (1839-1909) exploited this approach to isolate pure nickel via nickel tetracarbonyl (the Monde process).

harder and steel with up to 1.5% carbon can be used as cutting tools.

Iron and Steel Technology in the Current Era

The reduction of iron and copper likely were happening simultaneously. On the one hand it was easy to make copper and bronze and easy to fabricate bronze weapons and tools. But on the other hand, iron was found everywhere and any virtually any one could make iron anywhere. No expensive transportation and no expensive equipment. Just a lot of hard work.

The problems of making iron in large quantity revolved around getting the iron to melt. This was achieved with the "blast furnace." Conceptually, a blast furnace is not much different from the iron age device shown above, but it is a much larger scale (which helps with heat retention), and the air is provided by a mechanical compressor. The major innovation in the blast furnace is in *preheating the air*. The air is an unavoidable component of the process and if cool air has to be heated in the furnace and then is simply discharged as hot air, there is a limit to the maximum temperature that can be achieved. In a modern blast furnace, the air is pre-heated so that it does not cool the iron ore and carbon. For practical purposes, carbon is provided as "coke" rather than charcoal. Coke is obtained by heating coal to drive off volatile compounds containing C/H/O/N/S (coal tar).

The blast furnaces appeared to have been invented in China as early as the 1st Century of the current era based on the appearance of some cast iron used in trade. In Europe blast furnace technology does not appear until the 1400s in Belgium and 1500s in Britain. Presumably the air was provided by some wind or water-driven pump and the large scale of the furnace retained enough heat to liquify the iron, which could be drained off and cast to produce "pig iron." Pig iron[67] could be used by blacksmiths to make wrought iron and steel. The major advancements in iron production in Britain did not arrive until the 1700s when coke replaced charcoal[68] (Britain had lots of coal and dwindling forest). And the pre-heating of air was introduced in 1828 by James Beaumont Neilson.

Henry Bessemer (1813–1898) realized that steel could be produced directly from iron by burning more of the carbon out of the iron before it was cast.[69] Much of Bessemer's contribution was to see the value in a process and his ability to bring a product to market. Relying on successful experience he had in the glass industry, Bessemer's system (1856) involved taking the molten iron directly from a blast furnace and forcing air through it to reduce the carbon content. Although he patented the process and introduce it into several factories, it did not work

---

[67] *Pig iron* refers to the method of casting many small bars (piglets) from one main reservoir of molten iron (called the sow, i.e., mother pig).

[68] Abraham Darby (1709)

[69] In this, he followed and improved the methods of James Hall Nasmyth.

very well because it was a poor way to control the carbon content. Bessemer found the solution to precisely controlling carbon content in the work of Robert Forester Mushet. Mushet discovered that the only way to effectively control the low carbon content was to burn virtually all the carbon out of the iron and then add carefully measured quantities of a pig iron (Spiegeleisen), which is about 15% manganese and a known content of carbon to the melted iron. This approach gave Bessemer (1865) excellent control over the carbon content and allowed for careful alloying. The manganese was useful in removing phosphorus and other impurities into the silicate flux (slag) that floated on top of the molten steel.

Impact of Iron and Steel on Technology

The ability to make cast iron parts greatly impacted industry and set people entertaining inventions that had not previously been considered possible. The most important, I believe, was the pump. We have already mentioned that mining depended on dewatering mines. As water seeps in (from above or below), it needs to be removed so that mining can continue. In 1675, Samuel Moreland invented a piston pump that was very effective at lifting water out of mines, but at the time, driving the pump was limited to power sources available. Soon crude steam engines appeared (1698 by Thomas Savery; 1712 by Thomas Newcomen) and James Watt greatly improved the efficiency with a separate condenser (1775) and further improvements led to high pressure steam engines all facilitated by cast iron boilers and other iron parts. Soon steam engines were powering trains and ships. Trains in particular, created a new demand for iron in the form of rails and bridges. But iron rails wore out quickly and iron bridges needed to be massive to support the required weight.

The American Civil War (1861-65) brought new interest in improved artillery and the first ironclad ships (1862) proved that the days of wooden warships were over. Even before the war, there was talk of building a transcontinental railroad. The war proved the importance of rail traffic and seagoing steamships. After the war, the transcontinental railroad was built with the introduction of Bessemer steel manufactured in Pennsylvania by the Edgar Thomson Steel Works[70] (1874). The displacement of iron by steel allowed for more durable rails, and less massive bridges. Steel girders and steel-reinforced concrete created a boom in high rise buildings, which were never before possible.

Coal becomes an Important Fuel

Britain, which was destined to become a manufacturing leader in the 1800s, has no large forest. Thus, exploitation of coal as a source of heating began in the middle ages. Only rich people in Britain could afford wood, thus, poor people began gathering sea coal, which washed up on Britain's eastern shore from exposed underwater deposits. Unfortunately, the coal burned with much more smoke than wood and use of coal was discouraged (1306,

---

[70] Financed by Andrew Carnegie (1835-1919).

King Edward) until it was clear that expansion of industry was tied to this resource. In the 1800s, Britain's towns became smoky and air-polluted.[71] The invention of efficient de-watering pumps driven by steam engines facilitated the expansion of coal mining and the expansion of bituminous coal mining fueled the industrial revolution in Britain. The air pollution in Britain did not improve much until 1952 when a smog in London is credited with killing nearly 5,000 people. Coal soot (not carbon dioxide) has been credited by the US NASA[72] as contributing to the melting of glaciers in Europe by reducing the albedo of snow.

In the United States, coal was first found in northeastern Pennsylvania (1790) and it was hard anthracite coal, which produced much less smoke that coal imported to New England from Britain. Thus, exploitation of this coal to power the New England textile industry (1820s) did not cause the same degree of air pollution as encountered in Manchester, England. However, starting in the 1830s, bituminous coal from the Ohio River Valley has yielded both soot and sulfur dioxide (appearing as acid rain) falling down-wind on New England and the northeast United States.

---

[71] Charles Dickens (1852, *Bleak House*) "*Smoke lowering down from chimney pots, making a soft black drizzle, with flakes of soot in it as big as full grown flakes…Fog everywhere…in the eyes and throat…cruelly pinching the toes and fingers.*"

[72] Painter, TH et al. 1913. End of the Little Ice Age in the Alps forced by industrial black carbon. *PNAS*. 110(38):15216-15221.

5.4 Electrolysis of Metal Salts

In 1799 Alessandro Volta invented the electric pile in Italy. This system of an alternating stack of copper and zinc sheets separated by sheets of paper soaked with a conductor produced a continuous electrical current:

$$Zn \rightarrow Zn^{2+} + 2\ e\text{-} \rightarrow \text{ outside circuit} \rightarrow 2\ e\text{-} + 2\ H+ \rightarrow + H_2$$

Almost immediately, William Nicholson and Anthony Carlisle (1800) used the voltaic pile to break water into hydrogen and oxygen. This observation, which was consistent with the experiments in conservation of mass by Antoine Lavoisier a few years earlier (1778, 1783), clearly reinforced the atomic theory of matter and showed that the bonding of chemical elements had something to do with electricity.

Humphry Davy

Meanwhile in Britain, Humphry Davy saw an opportunity to attempt to isolate new elements from their minerals. He soon discovered that the current must be able to flow through the mineral to make the process work. His first attempts with potash (potassium oxide/hydroxide) failed until he added a small amount of water. This apparently provided an electrolyte pathway, which heated and melted the solid (mp 360°C). Once melted, the current could flow through the salt and potassium metal was soon produced. Sodium metal was also soon produced by a similar method. Davy continued to isolate a number of new elements including K, Na, Ba, Ca, B, Sr and Mg from various salts. The trick was to melt the salt so that it conducted electricity.

A number of other chemists followed Davy and many new metals were isolate. There were also spin-offs from this work in that potassium (isolated by electrolysis) could be used to displace other metals from salts. In particular, aluminum (Al) was reduced by this method in 1825 by Hans-Christian Ørsted and pure aluminum was isolated by Friedrich Wöhler in 1845. But this process was slow and expensive.

Aluminum

Aluminum (melting at only 660°C) proved to have many desirable properties (strong, light and shiny). In the mid-1800s, it was an exclusive substance that only the richest people could afford for applications such as knives and forks. There are anecdotes about the guests of Napoleon III of France (1848-52) eating off of gold plates, while the king ate off of aluminum plates. This monarch (1852-70) finance a cheaper process for producing aluminum using sodium rather than potassium in the hope of economically making new (light) armor for his army.

Aluminum could not be directly prepared by electrolysis because alumina (Al_2O_3) cannot be melted in any reasonable way (mp

2072°C).[73] Alumina is readily available as bauxite and can be separated from iron oxide impurities by extraction with concentrated base, filtration to remove iron oxide and reprecipitation with acid.

It seemed reasonable that if a lower melting mixture of alumina and some salt could be found, electrolysis might be possible. The problem was to find a salt that would dissolve in alumina and not react with metallic aluminum. Cryolite (Na_3AlF_6) has ideal properties as it melts at about 1000°C and dissolves alumina. With excess AlF_3 the melt temperature is reduced to about 950°C. The mixture is also less dense than molten aluminum. Thus, cryolite/alumina floats on top of molten aluminum and protects it from re-oxidation. This chemistry was developed in 1886 by Charles Martin Hall and Paul Héroult and it is called the Hall-Héroult process for electrolysis of alumina using carbon electrodes. At the graphite **cathode**:

$$Al^{3+} + 3\ e- \rightarrow Al$$

Fluoride (F-) is the only anion that is more difficult to oxidize than oxide (O^{2-}). Thus oxide (not fluoride) is oxidized at the fused-coke "sacrificial" **anode**:

$$2\ O^{2-} + C \rightarrow CO_2 + 4\ e-$$

The process normally runs at only 5 volts, but enormous amounts of electricity are consumed.

---

[73] Remember Coulomb's law. Here you have small, highly-charged ions and a very large lattice energy.

In practice, there are always some over-voltages, which cause incidental oxidation of fluoride to fluorine. This results in fluoride containing byproducts including $F_2C=O$, which must be trapped in a fume hood to protect workers and removed from the gases produced in the process before release to the atmosphere.

Relative to iron and steel, aluminum is still relatively expensive. Nonetheless, it appeared at a time (1886) when light, strong metals were essential to the advancement of several industries. In particular, the airplane was invented in the early 1900s and proved itself to have both civilian and military utility in World War I (1914-18). By the 1930s, wood and cloth were no longer the preferred materials for manufacturing airplanes.[74] Aluminum was worth its expense in the weight savings it provided.

Economics and Recycling

Although aluminum is expensive to make, its low melting point makes it cheap to recycle. More importantly, when a metal is recycled, the conditions of melting completely destroy any associated organic matter (e.g., pathogens) and the recycled

---

[74] However, it should be noted that the British navy had a wood and cloth biplane operating successfully off of its aircraft carriers throughout WWII; and because of a shortage of metals and metal workers, the British introduced a brilliantly state-of-the-art airplane (the de Havilland Mosquito) constructed from wood by cabinet makers. The advanced model had a maximum speed of over 400 miles per hour.

metal is as good as virgin material. Thus, aluminum is the perfect material to recycle and has excellent properties (strength, light weight, and corrosion resistance) desirable to a wide variety of structural and packaging uses. Overall, it only cost about 10% as much to recycle aluminum as it does to convert alumina into an equal mass of "virgin" aluminum.

We will examine recycling of other materials (below) and in contrast to the situation with aluminum, if you consider plastics such as polyethylene (which finds some of the same ended uses in packaging) you will find that (i) polyethylene cannot be returned to its starting material (ethylene) and (ii) melting polyethylene does not eliminate contaminating organic materials. Thus, most organic polymers cannot be recycled to *virgin-quality* products at any reasonable cost. In fact, most *hydrocarbon plastic materials* (polyolefins) would be most reasonably burned as fuel rather than recycled.

5.5 New Materials

Titanium

Titanium emerged after WWII (circa 1950) as a strategically important metal. It is less dense than steel and substantially stronger than aluminum. It was discovered in the late 1700s. But production of titanium metal is rather difficult because when heated with carbon, it forms titanium carbide (TiC) a hard and refractory material. It is typically isolated via $TiCl_4$ by various routs (displacement with Na or Mg).   The Kroll Process is used to make $TiCl_4$ from a common mineral ore:

$$FeTiO_3 + C \rightarrow Fe + TiO_2 + CO$$

$$TiO_2 + 2\,C + 2\,Cl_2 \rightarrow TiCl_4 + 2\,CO$$

Then

$$TiCl_4 + 2\,Mg \rightarrow Ti + MgCl_2 \ (800\text{-}850°C)$$

The result is a spongey mixture of Ti (mp 1668°C) and $MgCl_2$ which has to be mechanical broken up and leached to remove the magnesium salt. The particles of titanium can be melted in an arc furnace and cast into ingots, which are cooled under vacuum. This process is repeated to remove inclusions. Titanium has been used in military applications (submarines, high performance airplanes and as armor).

Titanium tetrachloride is useful as a catalyst in manufacture of polyolefins (see below).

Carbon

Since 2000, there has been much interest in synthesis of (covalently bound) allotropes of carbon related to graphite: graphene, carbon nanotubes, carbon spheres.

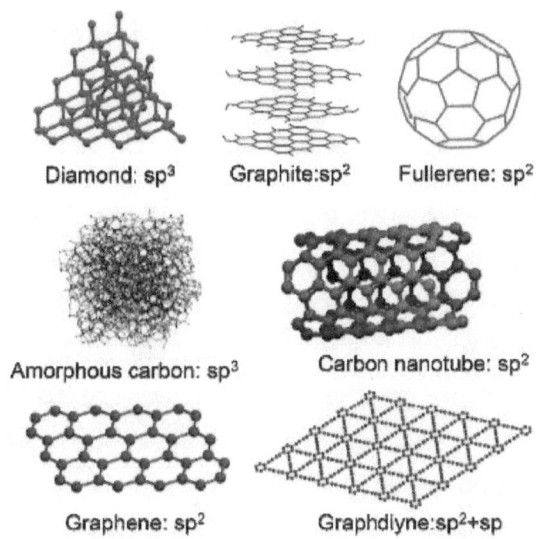

Author: Daniel Fishman; Source: https://www.quora.com

Carbon is an extremely light weight element (atomic mass 12 compared to aluminum 27, titanium 48, or iron 56) and has the potential to be stronger than steel and a better conductor. Thus, much research is currently underway with carbon.

5.6 Inorganic Silicate Polymers

Stone was the earliest durable building material used by humans. But, ironically, concrete was not widely available until the late 1800s. Similarly, glass was discovered early in our history, but we had little understanding of it until the last couple of hundred years.

Obviously, stone of various types (mainly sandstone and various forms of calcium carbonate, e.g., limestone and marble) have been quarried and used for construction since the Egyptians built the first pyramids (5000 years ago). By the time of the Greeks (3000 years ago), stone masons and sculptors had learned how to carve stone carefully and precisely. If you examine the Roman Colosseum (Coliseum) built 2000 years ago, some mortar was used in the construction. It was also used to make Roman aqueducts standing today.

But then, concrete disappeared until the 1800s. The technology was lost and almost all family dwellings were made of wood or stone. For example, in 1070, Winsor Castle (England) was built from wood, it was later rebuilt in stone. Depending on the period and location, stone structures were completed by filling gaps in the stones with either lime (calcium oxide), gypsum (calcium sulfate), bitumen, or a mixture of sand and clay. In the best old buildings, irregular stones were carefully shaped to precisely fit together. Why did concrete disappear for nearly 2,000 years?

The recipe for cement is actually very simple, if you have the correct ingredients. The Romans (300-450 CE) built structures from cement that are standing 2,000 years later. It was discovered (by some unknown engineer) that a mixture of lime (obtained by heating limestone):

$$CaCO_3 \text{ (limestone)} \rightarrow CaO \text{ (lime)} + CO_2$$

with pozzolana sand would harden into concrete when moistened. Pozzolana sand is a very fine volcanic sand found in Pozzuoli, Italy near Mt. Vesuvius which erupted in 79 CE. The ratios were 2- to 4-parts sand to 1-part lime. The secrete of Roman cement was lost for over a thousand years because the sand was not readily available. It was not until 1780 CE that the first modern record of cement was published: *Experiments and Observations Made With the View of Improving the Art of Composing and Applying Calcereous Cements and of Preparing Quicklime* by Bry Higgins. In 1793 CE, Eddystone Lighthouse in Cornwall, England was reconstructed using hydraulic cement (John Smeaton). The technology was advance by several patents in Britain and France in the early 1800s. Joseph Aspdin (1834) invented Portland cement (named after the quarry near Portland, England) by heating a natural mixture of chalk (a soft limestone) with fine silica clay in a lime kiln until it fused into a mass when all the carbon dioxide was driven off. The solid mass was finely ground and hardened quickly and evenly when water was added.

The issues with earlier cements were associated with the facts that (i) "clay" is a term that defines particle size, not chemical composition, (ii) limestone needs to be heated for a long time to drive off all the carbon dioxide, and (iii) limestone typically has various impurities. Any of the factors can reduce the quality of the product.

The chemistry of the cement reaction is rather straight forward.

When water is added to the mixture, the calcium oxide reacts to form calcium hydroxide, which dissolves to give a strongly basic solution. In the presence of strong base, silica sand (an insoluble polymer) is broken down into monomers, which are soluble.

$$(SiO_2)nOH + 2\ HO\text{-} \rightarrow (SiO_2)_{n\text{-}1}O\text{-} +\ HSiO_4^- + H_2O$$

This process is faster when fine particles of sand (i.e., large surface area to volume ratio) are used. As the process continues, the silicate monomers begin to form new polymers and the new polymers move through the sand and glue particles of sand together (i.e., the cement is starting to set). Eventually, a new polymer incorporating calcium ions is formed. The hardening process can actually continue for years.

Concrete is cement with pieces of silicate stone added such that larger volumes can be cast. Note that there is quite a bit of heat generated in the reaction and by including pieces of stone the stone both minimizes the heat per unit volume generated and absorbs some of the heat, which prevents some of the problems with cracking of large concrete pours through differential cooling.

Glass

Quartz (pure SiO_2) has a highly crystalline structure of silicate polymers which makes a three-dimensional array (mp 1700°C).

"Glass" is not a specific chemical composition; the term "glass" can be applied to any linear polymer (especially inorganic polymers) that become ridged with their chains in unstable configurations. Basically, the polymer is cooled faster than the chains can organize into the most stable crystalline arrangement. Here we are primarily concerned with glasses built from silicate polymers.

In these glasses, silica is the *former* and typically the major component (70 to 95%). *Fluxes* (such as sodium or potassium, carbonate) are added to facilitate melting and when melted they lose carbon dioxide to produce the corresponding oxides (0 to 16%). Glasses also incorporate *fixers* (calcium, magnesium, aluminum oxides) (1 to 5%) to make the glass impervious to water.

Special glasses can have lead oxide (to drastically change the refractive index) or colored metal ions to introduce color to the glass. As a general rule, higher levels of silica produce higher melting glasses. Introduction of boron oxide (B_2O_3) helps limit the thermal expansion and thus make glass less susceptible to cracking with a change in temperature (Pyrex).

Once a glass has solidified, it does not act like a liquid. It does not slowly flow. The polymer chains block random movements just as well as a crystal. Melting involves breaking the chains.

A product called "Bioglass" has a typical composition by weight: 45% SiO_2, 24.5% CaO, 24.5% Na_2O, and 6.0% P_2O_5. Bioglass, reacts with water and will bond to bone (which is built on calcium phosphate). It has applications in reconstruction of bones and teeth. It was developed by Larry L. Hench (1938-2015) in 1969.[75]

6.0 Chemistry of Non-Metals in the 1800s

6.1 Organic Molecules C/H/N/O

In Section 4.0 (above) we briefly considered the topic of organic molecules: shapes, functional groups and rudimentary nomenclature. Here we will build on that introduction.

Bioproducts

 Until recently (early 1800s) it was assumed that the chemicals of life were uniquely synthesized by living systems and since they all contained carbon, the term "organic" chemistry was originally intended to convey the idea that these compounds were only the products of living systems. This idea was gradually deposed in the early 1800s; and now while we marvel at the unique compounds (of carbon, hydrogen, oxygen, nitrogen etc.) that are formed by plants and animals, the "organic chemist" is well

---

[75] Hench L1, Jones JR. 2015. Bioactive Glasses: Frontiers and Challenges. *Front Bioeng Biotechnol.* 30(3):194.

equipped to make and modify such compounds and many that nature never attempted. Naturally occurring compounds can be lumped into several groups including (i) amino acids and their polyamides (peptides and proteins); (ii) carbohydrates (CH_2O starches, sugars); (iii) hydrocarbons (fats, oils, terpenes, steroids), (iv) other (especially nitrogen-containing compounds, purines, pyrimidines, quinones, porphyrins, alkaloids, etc.). Although some of these have been copied through "total synthesis," which was a popular theme of *academic* organic chemistry (e.g., 1940-2000)[76] and served the purpose of verifying structures deduced from spectroscopy and deconstruction; "spectroscopy" in its many forms and "fragmentation mass analysis" (a.k.a. mass spectroscopy) have progressed to the point where the value of "total synthesis" of natural products is somewhat reduced. Many of the reactions that were painstakingly developed in these academic achievements are not practical for mass production.

Coal

The field of synthetic organic chemistry received its first real boost when the metal industries shifted from charcoal (from

---

[76] If interested, consult the work of Robert B. Woodward, E. J. Corey et al. In my opinion: These chemists received tremendous accolades from their peers often for the clever work of their students, but contributed little to the economy. Unfortunately, they became the epitome of *academic chemistry*, which tended to confined organic chemistry to areas of decreasing interest. In particular, they distracted from the fields known as "molecular biology" and "materials science" which have blossomed.

wood) to coke (from coal) for manufacture of iron (Abraham Darby, 1709). But the history of coke goes back to the 1500s when a patent was issued to "purify pit-coal and free it from its offensive smell" so that it could be used in cooking and household heating (Dean of York, 1590). The offensive smell of coal was related to the presence of nitrogen and sulfur compounds, which produced a variety of volatile compounds when pyrolyzed. These were components of "coal tar and creosote," which originally only found applications in painting the hulls of wooden ships to prevent decay. In the 1800s, creosote also found use in treating railroad ties and utility poles to prevent decay.

But by the 1830s, when it was realized that "organic compounds" could be synthesized from inorganic compounds,[77] people began to separate the components of coal tar by fractional distillations and extraction. These components included aniline (1826 by Otto Unverdorben), indigo (1834 by Runge[78]), thiophene (1883), benzene (1825 by Michael Faraday), phenol (1834), naphthalene (1821 by John Kidd), quinoline (1834 by Runge) and methylpyridine (i.e., picoline, 1846 by Anderson). Of these, benzene was one of the largest components. Britain and Germany had large sources of coal; and, thus, could produce

---

[77] Principally by Justus von Liebig (1803-!873).

[78] Friedlieb Runge (1794-1867) isolated caffeine from coffee beans in 1819.

large quantities of coke and hydrocarbons benzene, toluene, ethylbenzene and xylenes.

The Synthetic Dye Industry

A variety of natural compounds had been obtained from ancient times and used to color silk, wool and cotton. Synthetic dyes were about to completely change this picture.

Aniline was synthesized from benzene by nitration to nitrobenzene and reduction of the nitro group to and amino group by Nikolay Zinin in 1842. August von Hoffmann, [79] who had also worked on the composition of aniline from coal tar and indigo, established a research group. One of the important topics of the day was quinine, which the British foreign service personnel consumed as "gin and tonic."

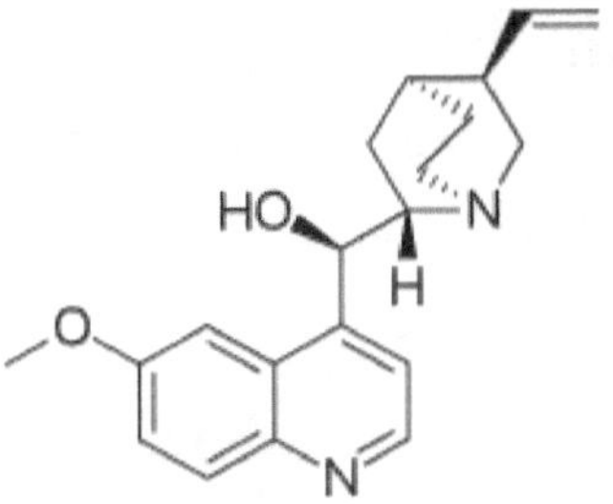

Quinine

Author: Vaccinationist; Source: Wikimedia Commons

---

[79] Zinin and August von Hofmann were students of von Liebig. Hofmann became the director of the Royal College of Chemistry in London (1845).

The European nations were busily claiming unorganized territories (principally in the tropics of Africa and East Indies) as "colonies." Except for South Africa and Australia, these "colonies" were little more than military outpost where the natives were forced to produce natural resources for export to Europe. The governors, soldiers and bureaucrats that oversaw these outposts were very susceptible to tropical diseases (which we will discuss in Part III) and the only useful medicine was quinine, which was in short supply. Of course, the structure of quinine was unknown; but it seemed worth the effort to see if aniline could be converted to quinine.

The young William Henry Perkin (1838-1907) came to study coal tar chemistry with von Hofmann in 1853. Perkins was apparently very smart, but he was not devoted to science/chemistry. He came from a very practical background where making money was the measure of success, not publishing scientific papers. Thus, when "luck" dropped an unexpected discovery in his lap, he recognized the economic implications and moved directly to cash in.

According to accepted history, after receiving the basic courses in chemistry, Hofmann assigned Perkin the task of making amino derivatives of anthracene, naphthalene and benzene. These would be routine applications of Zinin's methodology of nitration and reduction. And it is quite possible that other students had made some of the same observations that Perkin was about to make. The accepted story is that rather than working in Hofmann's well-equipped laboratory, Perkin moved his research to his apartment in London over the Easter holiday in 1856. There he is said to have attempted a reaction with aniline, toluidine and potassium dichromate, which produced a tar. Perkin soon realized that the extract of the black tar with ethanol gave a purple solution. In follow up experiments in his home (not revealed to Hofmann), the purple solution durably dyed cloth (especially silk) purple.[80]

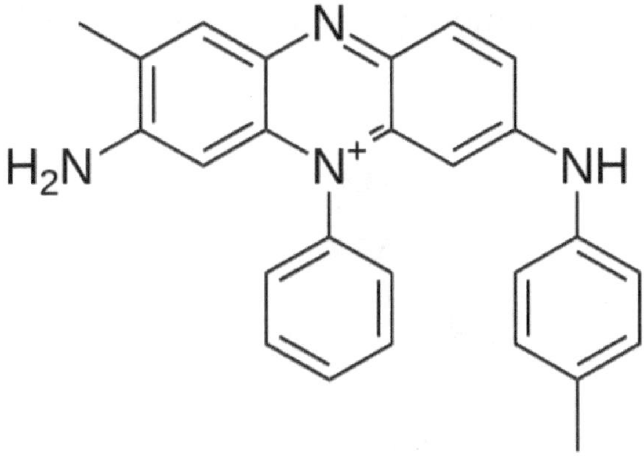

"Mauve"

Author: Reubot; Source: Wikimedia Commons

---

[80] I do not believe this story for a minute; but you can believe what your wish. I think the key observations were made in Hofmann's lab and the story was created to give Perkin independent inventorship and patent rights (filed 26 August 1856). Regardless, Hofmann did not seem to care much as the reaction added little to science.

See:
http://www.colorantshistory.org/HistoryInternationalDyeIndustry.html

Realizing that this could be valuable, Perkin resigned from Hofmann's laboratory and pursued a career as a dye manufacturer.

Perkin filed a patent dated 26 August 1856. By January of 1857, a silk dyer (Thomas Keith) was impressed with the product and placed an order. Perkins and his father and brother established a factory on the outskirts of London in late 1857 and they were in business. It took two more years to get the product adapted to dying cotton.

A.D. 1856 . . . . N° 1984.

Dyeing Fabrics.

LETTERS PATENT to William Henry Perkin, of King David Fort, in the Parish of Saint George in the East, in the County of Middlesex, Chemist, for the Invention of " PRODUCING A NEW COLORING MATTER FOR DYEING WITH A LILAC OR PURPLE COLOR STUFFS OF SILK, COTTON, WOOL, OR OTHER MATERIALS."

Sealed the 20th February 1857, and dated the 26th August 1856.

Perkin would not license his patent and other companies began experimenting with similar reactions. Interestingly, von Hoffman and François-Emmanuel Verguin soon independently discovered that under similar oxidizing conditions aniline and carbon tetrachloride produces "red fuchsine"

"red fuchsine"

Author: Edgar181; Source Wikimedia Commons

Again, von Hoffman appears to have made no effort to patent; while Verguin did.

These are interesting reactions but (especially since the structures of the several products were not known at the time) they offered no systematic approach to making dyes. Nonetheless, out of the experimentation with aniline (which was prompted by the easy money coming from the manufacture of dyes) a general method of synthesizing compounds with extended pi-electron clouds was discovered.

The Quantum Mechanics of Dyes

As we have seen in Section 1.4, the wavelengths of light that we can see range from 350 to 700 nm. So, in order to be a dye or colored compound, a molecule must absorb light in that range. Here is where quantomechanical theory and the practice of money making (we call it business) come together. Schrodinger applied his equation to a number of physical models. The simplest is called the particle in a box (i.e., "an infinite potential well" as the physicists might say).

$$\Delta E = hc\ /\ \lambda$$

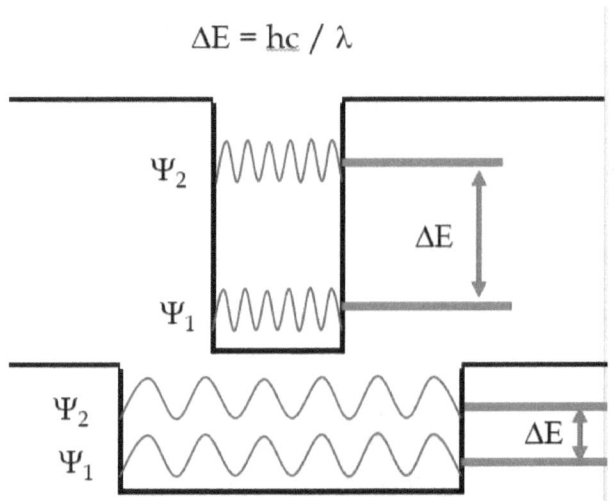

Basically, the smaller the box (or narrower the well) in which the electron is confined, the greater the energy separations between the wave functions ($\Psi_1 \rightarrow \Psi_2$). Imagine the pi-electrons of an ethene molecule ($H_2C=CH_2$) contained in a small "box" provided by the pi-bond orbital. The energy-level separations are rather large: The first excitation energy is 185 nm. If we make a di-ene ($H_2C=CH-HC=CH_2$), then the box confining the p-electrons is bigger and then energy level separation is less (and the wave length of absorption is longer, 215 nm). Following this trend, vitamin A (5 adjacent double bonds) absorbs at 328 nm.

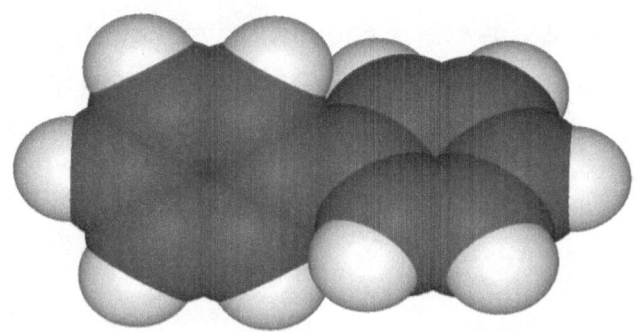

Vitamin A

Author: NEUROtiker; Source: Wikimedia Commons

Benzene absorbs at about 260 nm; naphthalene up to 285 nm, and anthracene up to nm 380 nm.

Author: Edgar181; Source: Wikimedia Commons

Note that biphenyl absorbs at about the same place that benzene absorbs because the rings are not co-planar (i.e., the electron box is cut in half by rotation around the single bond between the two rings). All of these compounds appear colorless to us. On the other hand, carotene (eleven double bonds adjacent to one another) absorbs at about 450 nm (400-500 nm) and we see it as an orange color. Of course, graphite is black because there are many long pathways that can absorb any wavelength; meanwhile diamond is colorless, because there are no p-electrons free to move around the molecules.

Diazonium Coupling

Von Hofmann's lab continued to work on aniline compounds and in the spring of 1858, his associate Peter Griess discovered that reaction of *aromatic* amines with sodium nitrite ($NaNO_2$) in hydrochloric acid an around 0°C (ice water) produced an intermediate that was stable enough to work with (under these conditions, products from *aliphatic* amines almost immediately decompose).[81] This intermediate had a positive charge and is represented as

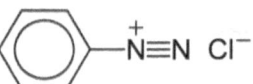

Diazonium salt of benzene (from aniline)

Author: Benjah-bmm27; Source: Wikimedia Commons

---

[81] Peter Griess (1858) "*Vorläufige Notiz über die Einwirkung von salpetriger Säure auf Amidinitro- und Aminitrophenylsäure*," (Preliminary notice of the reaction of nitrous acid with picramic acid and aminonitrophenol), *Annalen der Chemie und Pharmacie*, 106 : 123-125.

Griess and von Hofmann spent several years making and studying the chemistry of these diazonium compounds (1858-1862). One of their coworkers was Carl Alexander Martius (1838-1920) who (like Perkin) made discoveries he did not share.

Bismarck Brown

Author: Smokefoot; Source: Wikimedia Commons

In 1863, Martius apparently made 1,3-dinitrobenzene, reduced it to the diamine and was trying to diazotize 1,3-diaminobenzene. The product was a brown colored compound, which was an effective dye (Bismarck Brown). Martius separated from the Hofmann group and began a career as a dye manufacturer in collaboration with Heinrich Caro (1834-1910) at the Roberts, Dale & Co. in Manchester England (the center of the cotton textile industry). Caro was primarily involved with printing black on cotton and was is competition with John Lightfoot who developed a process for printing aniline onto cotton and then converting this to a dark dye *in situ*.

The structures of these compounds were not known; but in 1865, Professor Friedrich August Kekulé (1829-1896) proposed a structure of benzene that made the chemistry of aniline fall into place. He had realized that carbon was tetravalent in 1857 and proposed the cyclic structure of benzene at a time when there was no known physical basis for the structure. Extension of Kekulé's ideas to nitrogen and oxygen had a profound effect on organic chemistry in the latter 1800s.

This work is in the public domain in the US

Martius and Caro synthesized a number of colored compounds in 1863 -1864 including aniline yellow. The mechanism of the coupling reaction was diazo coupling (electrophilic attack on benzene), which was found to be a general reaction of a diazonium with aromatic rings that were activated by the presence of a -NH$_2$ (aniline) or -OH (phenol) substituent:

Author: Ben Mills; Source: Wikimedia Commons

The azo (-N=N-) linked aromatic systems now had long pathways for the p-electrons to move in; and, thus, absorbed in the visible spectrum. This invention provided a great incentive for chemists to generate more derivative of benzene, naphthalene and anthracene. A great many aromatic compounds were synthesized and many were converted to the corresponding azo-dyes. We will see later that this body of knowledge paid off when scientists learned the microbial basis of many diseases in the 1890s.

Heinrich Caro

Heinrich Caro was a practical business man, not a great chemist. His career started as a chemist in a cotton printing firm and he became familiar with the business and legal aspects of the chemistry companies. He worked briefly for several well-known chemists (e.g., Victor Meyer's father, Robert Bunsen) before joining *Chemische Fabrik Dyckerhoff Clemm & Co* (later BASF) about the time that structural organic chemistry took hold. He was involved with the isolation of acridine from coal tar (1870). He worked with Adolf von Baeyer (1835–1917) when Baeyer synthesized indigo dyes (1878). But Caro's real strengths were in patenting inventions and he led BASF's patenting of alizarin dye.

Patents

Patents are a form of "intellectual property," i.e., knowledge that has value and is owned by an inventor (or assignee). The rationale for a patent is that the government of a country is motivated to stimulate technological advancement. In exchange for an inventor divulging certain knowledge (in a form and detail that others "skilled in the art" can replicate the invention), the government grants the inventor a period of exclusive legal usage (ownership) of the knowledge as divulged and claimed in the patent. The inventor may sue people (infringers) who use the knowledge without permission of the inventor in court and collect financial damages, but the government does not defend the inventor's interest.

The use of patents in Europe began in the 1300 and 1400s. The process of obtaining a patent has varied over time and from country to country. In Europe, because the countries are small, obtaining patents for the same invention in many different countries is generally required for the patent to have any value. Because the economy of the United States is quite large today, an American patent is more valuable.

The review of a patent can be complex. The government must ensure that the invention is significant (i.e., novel and not and obvious extension of existing knowledge). Moreover, the government must ensure that the rights being granted have not already been given to some other inventor. Thus, a patent search and broad literature search are required to satisfy a patent examiner that the invention is available for patenting and that it has not already be granted to someone else. Because this process can take time (during which the patent is not granted), the priority of the claim must be considered if people have overlapping claims.

It must be noted that in businesses where research activities are typically the basis for founding new business or products (consider the example of Perkin), employees are generally required to sign contracts with employers when they are hired. These contracts require the employee to agree to (i) non-disclosure, (ii) non-compete, and (iii) intellectual property rights assigned to the company. The examples of Perkin and Martius described above point out the importance of such conditions of employment. Academia has historically been less rigid. But as universities have realized the value of inventions, most faculty members are bound to their universities by more or less the same rules although the universities are less prone to enforce these conditions. Students (e.g., graduate students) are more problematic. They generally do not have a contractual arrangement with the school or the faculty. But there are "unwritten rules" that can undermine a student's career if broken.

There has been a relatively recent change in the patent law of the United States. For many years, companies kept detailed and rigidly documented notes (daily signed and dated by supervisors) regarding experiments and ideas generated by its employees. The reason was that the law gave priority to the *date of invention*. The new law has changed to a "first to file" priority system because litigation of the date of invention is very complicated and subject to ambiguity. In contrast, the date/time of filing is very clear.

In the event of intellectual property owned by a private party in two countries that go to war with one another, there is a general government office called the *alien property custodian*. Basically, the intellectual property (indeed all property) owned by a company in a country formally "at war" with the country of residence of the owner is placed at the disposal of the alien property office until the war is resolved to ensure that the technology is not used against the interest of the country. When the war is resolved, the private organizations can reclaim their property in many cases. We will see cases (below) where corporations were caught in awkward legal and public relations (political) situations because of their international business interests in countries that went to war with one another.

Phenol, Salicylate, Aspirin and Bayer

Form antiquity, it had been known that willow bark contained an ingredient that relieved fever and pain. In 1763, the British Royal Society published a report confirming the medicinal use of this material. In 1828, Joseph Buchner isolated the active ingredient (salicin).[82]   In 1838, Raffaele Piria (1814–1865) hydrolyzed salicin to remove the sugar and oxidized the other component (2-hydroxymethylphenol) to salicylic acid. Salicylic acid proved to be more effective at pain relief than salicin.

In 1853, Charles Gerhardt (1816-1856) was making anhydrides of many acids by reaction of their salts with acetyl chloride (CH_3COCl) and applied his technique to a salt of salicylic acid. He *assumed* the product he isolated in 1853 was an anhydride and he did not follow up his work. In 1859, von Gilm did essentially the same thing with the acid itself. It was not until 1869, that the products made by Gerhardt and von Gilm were shown to be the

---

[82] Some sources give Leroux in 1929 or 1930. (no author). Salicin, Salicylic Acid and the Salicylates. *Ind Med Gaz*. 1878 Sep 2;13(9):248-252

same thing and that they were the acetyl ester of the phenol, not the anhydride.

Through the American Civil War (1861-65) gangrene turned many survivable wounds of the arms and legs into death sentences. At the time, the nature of infection was not understood. The connection between microorganisms and disease was only slowly growing in the 1860s. Nonetheless, in part from Louis Pasture's demonstration that beer and wine did not go bad without exposure to sources of infection, the idea that it might be able to prevent or overcome infections by chemical means gained some ground. In 1867, Joseph Lister (1827–1912) started applying carbolic acid (phenol) to wounds to prevent infection and as a method of sterilizing surgical instruments. His success attracted substantial attention by 1873. By this time, the corrosive effects of phenol on skin was recognized and the idea came to imminent chemist Herman Kolbe (1818-1884) that the compound that he and one of his students (Rudolph Schmitt, 1830-1898) had synthesized in 1863 (salicylic acid), might be less objectionable than phenol while retaining the anti-microbial effects.

The Kolbe-Schmitt Reaction (1863)

Author: Edgar181; Source Wikimedia Commons

He did some experiments with salicylate and successfully prevented fermentations. He also found that, salicylate could be tolerated internally and that it seemed to have antibiotic effects because it reduced fever and pain.

Meanwhile, Friedrich von Heyden (1838-1926) received his doctorate as a student of Schmitt in 1874. Von Heyden set up a factory to manufacture synthetic salicylate. In clinical trials their product proved to be very popular for pain and fever (1876), although it was found not to be an effective antibiotic. At this time, salicin was recognized to be less irritating to the throat and stomach than salicylic acid. [83]

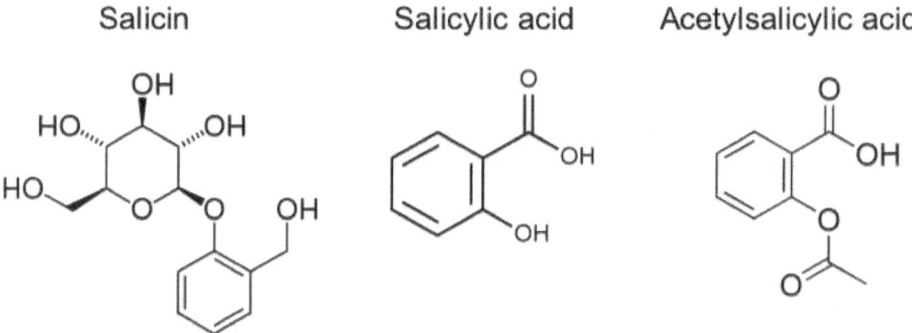

Salicin   Salicylic acid   Acetylsalicylic acid

Author: Andrea Wills (aewills@stanford.edu);

Source: https://abouquetfrommendel.wordpress.com/

---

[83] T. Maclagan. The Treatment of Rheumatism by Salicin and Salicylic Acid. *Br Med J.* 1876. 1(803): 627.

In a different venue, Felix Hoffman (1868-1946), working for the Bayer[84] Company under the direction of Arthur Eichengrun, synthesized acetylsalicylate in 1897. In this process, it was clarified that the acetyl group was on the phenol (forming an ester); i.e., not on the carboxylate forming an anhydride (and not on the aromatic ring). The motivation for the work was to find a less irritating material than salicylate from the von Heyden Company. Acetylsalicylate is much better tolerated than salicylate itself[85]; and under the trade name Asprin®, they sold their product around the world by 1899.

In 1901, the von Heyden Company also began selling acetylsalicylate in Germany (where there was no patent) and in other countries where Bayer had obtained patents. Obviously, litigation followed and it was resolved that the Bayer patents had been incorrectly issued because the work of Gerhardt, Kolbe and others (before 1870) had been overlooked.[86]

It is worth mentioning that during WWI (1914-18) production of acetylsalicylate was limited by the availability of phenol. Thomas Edison had set up a phenol plant to make phenol for

---

[84] Be careful not to confuse Bayer (i.e., Friedrich Bayer (1825–1880)) with Baeyer (Adolf von Baeyer [pronounced "Buyer" in English] (1835–1917)).

[85] Presumably because of the slow hydrolysis of the ester releasing the phenol.

[86] The reader should note that some of the popular literature has been slanted to favor one party of the other.

Bakelite (see below) phonograph records, and *while the U.S. was still neutral in the war*, Germany arranged to buy all of Edison's excess phenol (at a very high price) and route it to their salicylate-manufacturing subsidiary in the U.S.; from there the salicylic acid was shipped to Germany. This was all legal, but (because the German and British phenol shortage was caused by the use of phenol to make picric acid (trinitrophenol) explosive, see below) the direct shipment of phenol to Germany would have raised eyebrows. This became known as the "phenol plot;" and when it was made public, Edison sent his excess phenol to the U.S. military.

When the U.S. entered WWI in 1917, German property in the U.S. became subject to the "alien property custodian" who was responsible to see that it was not used to support the enemy nation. Ultimately, Bayer's company (including the intellectual property, trademarks, brand names, etc.) was seized and auctioned to an American company. The valuable marketing tools were ultimately recovered by Bayer for approximately a billion dollars in 1994.

6.2 Inorganic Materials of Non-Metals

Today, we tend to take for granted the availability of mineral acids (chloride, sulfate, nitrate) and ammonia (NH_3). But these were not always available and their introduction was critical to advancements in civilization.

Oil of Vitriol and the Mineral Acids

Elemental sulfur (brimstone) occurs naturally in volcanic regions and large deposits on Sicily were undoubtedly known from early times. The burning of sulfur for religious and medicinal purposes was well known. However, it was not clearly identified as an element until 1777 (by Lavoisier).

As noted above, most metals of antiquity were either found uncombined (like gold) or were roasted from their sulfides. But along with the natural deposits of insoluble metal sulfides, biological oxidation had created metal sulfates, which were soluble and which percolated into the soil where they were deposited as hydrated metal sulfates ($MSO_4.X(H_2O)$). These compounds were typically colorful and were termed "vitriol," more specifically, *green vitriol* was a hydrate of iron sulfate and *blue vitriol* was a hydrate of copper sulfate. However, magnesium sulfate (Epson slat), calcium sulfate (gypsum) and aluminum sulfate[87] (alum) are essentially colorless and were called "white vitriol." These were certainly recognized concurrent with the earliest mining activities and as formations in caves. Because they are soluble in water, they could be extracted from the earth in which they were found with boiling water, which upon cooling precipitated interesting crystals of

----

[87] Aluminum sulfate has a different chemical formula and different sources, but its chemical properties and processing are similar.

vitriol. There is evidence that these compounds were routinely produced for various applications between 300 BCE and 300 CE.

There is evidence that the Arab scientist Jābir ibn Hayyān (est. 721–815 CE) may have discovered some of this chemistry, but around 1600[88], alchemic recipes in Western Europe started appearing that seem to describe the dry distillation of various vitriol compounds. When this is done, the metal oxide (MO) tends to separate as a solid leaving H_2SO_4 (sulfuric acid), which distills at 337°C (with decomposition to H_2O and SO_3). If the vitriol contains elemental sulfur, which is not unusual for mineral ores, the sulfur (S_8) will decompose to a dimer as it distills and the dimers recombine as poly-sulfides that are orange to red. These early descriptions frequently mention "red oils," which became known as "oil of vitriol." Indeed, (anhydrous or very concentrated) sulfuric acid is oily, i.e., much more viscous and denser than water.

$$CuSO_4 \bullet H_2O \rightarrow CuO + H_2SO_4$$

However, it was not long before a more practical way of producing sulfuric acid was found. Johann Glauber (1604 (?)–1670) burned sulfur with saltpeter (perhaps he was studying the chemistry of gun powder) and produced sulfur trioxide, which dissolves in water to produce sulfuring acid. It is relevant that burning sulfur in air (which was likely done frequently by alchemists) does not produce sulfur trioxide, but rather sulfur

---

[88] Johann Thölde (1565–1624) a.k.a. Basil Valentine may have produced a mixture of nitric and hydrochloric acids called *aqua regia*, which has the property of dissolving gold and platinum.

dioxide (SO_2), which forms a much less stable hydrate (H_2SO_3) that decomposes to sulfur dioxide if you attempt to isolate it. Joshua Ward (1685–1761) used Glauber's method in the first commercial manufacture of sulfuric acid (1736). This was soon displaced by John Roebuck's (1718–1794) "lead chamber" process. This invention was somewhat serendipitous. The key to Glauber's oxidation was that nitrate breaks down to NO (nitric oxide)[89], which is oxidized by air to ($\bullet NO_2$) nitrogen dioxide (with an unpaired electron). Nitrogen dioxide is able to oxidize sulfur dioxide to sulfur trioxide as it is returned to nitric oxide. Originally, Roebuck's chemistry was intended to be the same as Glauber's chemistry with the lead-lined (wooden) condensing chamber merely a cheaper fabrication than glass for the isolation of sulfuric acid on a large scale. But the design had the fortuitous effect of recovering the nitric oxide (NO) (i.e., it was not allowed to escape into the air), which could then be cycled catalytically in the lead chamber.

$$NO + O_2 \rightarrow NO_2$$

$$NO_2 + SO_2 \rightarrow SO_3 + NO$$

(Thus, reducing the need for saltpeter.) Over the 200-year that this process has been used, various sources for both sulfur

---

[89] The nomenclature is a little confusing because **nitrous oxide is N_2O** (laughing gas). Humphry Davy (1778–1829) was working at The Pneumatic Institution in Bristol, which was devoted to medical experimentation with various "airs" (e.g., nitrous oxide, N_2O). In 1799, Davy discovered how to purify nitrous oxide (removing toxic NO and NO_2) and acquired the habit of liberal use of "laughing gas."

dioxide and nitrogen dioxide have been developed. In the most recent versions, the sulfur dioxide was provided as a waste material from roasting sulfide ores and the sulfuric acid produced as a byproduct of what would now be called "green chemistry" (designing chemical processes to minimize waste and toxic materials).

Glauber used sulfuric acid to produce nitric acid (by reaction with saltpeter followed by distillation of the nitric acid/water azeotrope (68% HNO_3, b.p. 121°C)) and "marine" or "muriatic" acid, known to chemist as hydrochloric acid (HCl) by treating sea water with sulfuric acid and distilling the hydrochloric acid/water azeotrope (20% HCl, b.p. 109°C). He also went forward and reacted a number of metal oxides with these acids to produce nitrate and chloride salts. Some of the chloride salts can be distilled (e.g., $AsCl_3$, $SbCl_3$, and $ZnCl_2$).

When this approach what attempted with manganese dioxide (mineral pyrolusite) by Jan Baptist van Helmont (1630), the manganese(iv) was reduced to Mn(ii) and chloride anion was oxidized to chlorine gas. This approach was used by Carl Wilhelm Scheele (1774) to isolate chlorine and characterize it for the first time.

The Halogens

Chlorine (Cl_2) was available in unlimited quantity after Davy (1810) discovered that it could be obtained by electrolysis of a solution of table salt.

The Napoleonic Wars were in full swing and the armies required potassium nitrate of gun powder. The production of potassium utilized sodium carbonate (see below), which was isolated by combusting seaweed to "soda ash" much the same was that "pot ash" is obtained from wood. Soda ash was in fact the material that Davy had used to isolate sodium by electrolysis, as he had isolated potassium from potash.

Of course, the same process was employed in France. For reasons not entirely clear, Bernard Courtois (1777– 1838) treated seaweed residue with excess sulfuric acid at high temperature and apparently managed to break down some of the organic compounds[90] that accumulate iodine from seawater. Iodine vapor sublimed out and was condenses as crystals on cool glass surfaces. He knew he had something new and sent samples to a number of leading scientists. Davy appears to be the first to identify iodine as an element (1811).

Carl Jacob Löwig and Antoine Balard independently discovered bromine (1825/26) by chlorinating mineral waters and extracting the bromine with organic solvent. When the solvent was evaporated the brown liquid remained.

Today, bromine and iodine are typically extracted from brine solutions or sea water by chlorination.

---

[90] 3-iodo-L-tyrosine, 3,5-diiodo-L-tyrosine, 4-iodophenol, and 2-iodobenzoic acid have been identified. Yang Y et al. Selective Identification of Organic Iodine Compounds Using Liquid Chromatography-High Resolution Mass Spectrometry. *Anal Chem.* 2016. 88(2):1275-80.

$$Cl_2 + 2\, Br^- \rightarrow Br_2 + 2\, Cl^-$$

Iodine is collected incidental to the recovery of potassium nitrate in Chile and Peru.

Ammonia and Nitric Acid

The modern history of ammonia (NH_3) is very interesting. It begins in the 1700 and 1800s when farmers realized that soil became "worn out" after several years of intense agriculture. Actually, it had been known for some time that manure from farm animals and people could maintain a level of productivity of agricultural soil and in some societies inedible/undesirable portions of fish were/are added to soil. But it was not until the 1700s that the population in Europe reached levels (after recovery from the Black Death) where food production was limited by the need to replace nutrients in the soil. The tendency in British North American colonies was to simply clear new land for farming when the productivity of soil dropped off. That luxury did not exist in Europe.

At the time, the carbon found in plants was assumed to come from the soil and it was argued that fertilizer merely facilitated the breakdown of humus to a form of carbon that plants could absorb. Nicolas-Théodore de Saussure (1767–1845) studied photosynthesis and realized that the carbon was coming from the air (as CO_2) and oxygen released by plants was likely coming from water (not CO_2).

During the Napoleonic Wars (1803-15), the French obtained saltpeter (potassium nitrate) by composting manure and urine in especially designed sheds called niter-beds. Once the biological oxidation was deemed complete, the mass was extracted with hot water. The leachate contained nitrates and chlorides of sodium, magnesium and calcium. To remove the magnesium and calcium and ensure that there was as much potassium ion as nitrate ion present, it was treated with potash (wood ash, KOH). Soda ash from seaweed (containing sodium carbonate) was added to precipitate magnesium and calcium as carbonates and/or hydroxides leaving behind a solution containing organic materials, potassium, sodium, nitrate and chloride. After removing the magnesium and calcium precipitates, the remaining solution was concentrated and glue could be added to coagulate the organics that collected on the surface, from which they were skimmed. The final solution was concentrated until sodium (and perhaps potassium) chloride precipitated from the boiling liquid (NaCl 0.39 g/mL and KNO_3 0.24 g/mL at 20°C). Then a series of fractional recrystallizations were used to ultimately yield saltpeter pure enough for gunpowder (NaCl 0.36 g/mL and KNO_3 0.01 g/mL at 100°C). In the mid-1800s, European powers began looking for mineral sources of nitrates and found some in South America.

Justus Liebig (1803–1873) confirmed de Saussure's observations and began studies of the effects of various minerals (Ca, K and Fe) on plant growth. He soon found that nitrogen was also an essential element and realized nitrate that was the primary benefit of manure. Moreover, he realized that (as Carl Sprengel

(1787–1859) had claimed) you cannot compensate for a shortage of one nutrient with an over-abundance of another. The limiting nutrient will determine the productivity of the plants.

It turns out that most nitrate in the soil is produced by high-energy lightning strikes in the air. At exceptional high temperature, nitrogen and oxygen molecules are ionized/radicalized and recombined:

$$N_2 + 2\ O_2 \rightarrow 2 \bullet NO_2 \text{ (brown)} \leftrightarrows N_2O_4 \text{ (colorless)}$$

Notice that this reaction alone would ensure that oxygen would be depleted from the primordial atmosphere and delay the accumulation of oxygen in the atmosphere until the green plants covered most of the earth surface.[91] Further reaction with water and oxygen, provide a dilute nitric/nitrous acid in rain.

All nitrate salts are soluble; thus, nitrates were washed into the oceans and supported the evolution of life in the sea as a source of energy and nitrogen for essential biomolecules (DNA, RNA, protein). It follows that the productivity of the soil has always been limited by the lack of nitrogen.

Nitrogen in the marine environment followed the food chain to fish and from fish to birds. Birds then deposited guano on the land. But most nitrate from bird (and bat) guano is simply recycled into the oceans and aquatic systems. The only places that nitrate is retained in the soil is in caves (where birds and bats

---

[91] Nitrogen dioxide has an unpaired electron that causes it to have a brown color. It reacts with itself to form N_2O_4 which is colorless.

roost)[92] and on high/dry plateaus, which explained the discover of nitrate deposits of economic importance in the desert shared by Chile, Peru and Bolivia west of the Andes mountains. Apparently, for millions of year birds fed off of the fish that school in the Peru Current and roosted on highlands along the coast. Part of this land drains into a basin to the east where the nitrate deposits are found on the western fringe of the Atacama Desert. These deposits were used locally to extract nitrate for gunpowder during the time of Spanish control.

Once Liebig's principles of plant nutrition were circulated, it was clear that saltpeter (potassium nitrate) production would be split between military and industrial applications (explosives) and agricultural (fertilizer) applications. Thus, when a sample of Peruvian potassium nitrate was sent to Spain in 1821 by Pedro Fuentes it was certain to soon find a market, and it did in in England (1827), the US (1830), and France (1831).[93]

The Europeans (especially the British) began to exploit the mineral deposits creating economic rivalry among the South American governments and resulting in a war between Chile (backed by Britain) and Peru/Bolivia (1879). Chile won and

---

[92] Cave guano (as well as human waste) have been collected in the United States during the war of independence and the American civil war for use in making gun powder.

[93] *Memoirs of the Peruvian wise man Mr. Mariano Eduardo Rivero* contained in *EL SALITRE (NITRATE), Historical Resume since its Discovery and Exploitation.* Roberto Hernández C., Valparaíso, Fisher Hnos., 1930.

annexed substantial parts of Peru and Bolivia then and later. In the short term, Europe had a ready supply of nitrate, but the Americans also began importing nitrate from Chili. As the western populations began to expand rapidly with the understanding of the cause of disease (1870) and improved food supply, British scientist William Crookes (1832-1919) did a simple calculation based on the known nitrate resource. In 1898, Crooks estimated that within 20 to 30 years the population of Europe would exceed the agricultural potential of the Chilean nitrate and that famine (not seen since the 1840s in Ireland) would follow. Crookes identified air as the obvious source of nitrogen that must be tapped for long term agricultural success.

This sent many minds to working. Several groups attempted to mimic nature with electric sparks through air. In the United States Thomas Edison (1847-1931; direct current) and George Westinghouse (1846-1914; alternating current) had battled to provide an economical way to produce and distribute electricity and take advantage of Edison's light bulb (1879). Westinghouse eventually won out and alternating current generators began to pop up everywhere. William Armstrong probably built the first hydroelectric plant in Northumberland in 1878 to power an arc lamp, but this was a minor part of his arms empire, which build breech-loading rifled cannons and warships at Newcastle. With the spread of the electric light bulb, there were about 200 hydroelectric plants in the U.S. by 1889.

In 1901 Charles S. Bradley and D. Ross Lovejoy patented a process for oxidizing atmospheric nitrogen in an electric arc.

They formed a company (the Atmospheric Products Co.), raised capital and built a plant at Niagara Falls, NY[94] (where hydroelectric power was cheapest) to make nitric acid. The process was not particularly efficient and they went out of business by 1904. However, the idea was still attractive to Samuel Eyde in Norway because of an abundance of hydroelectricity in Norway and the less secure access to mineral nitrates in Europe.

Samuel Eyde (1866 – 1940) was from a wealthy Norwegian family and obtained a degree in engineering in 1891 in Berlin. Moving back to Sweden, he established the largest engineering firm in Scandinavia with offices in Stockholm and Kristiania (after 1925, Oslo). Meanwhile, artic explore and inventor Kristian Birkeland had done work on what we would now call a "rail gun" (i.e., projectile accelerated by a magnetic field), which had recently failed as a result of a short circuit creating an enormous electric arc. Birkeland and Eyde met socially in 1903, when Eyde mentioned that he needed a gigantic electric spark to make nitrate from air, Birkeland knew just how to do that. Eyde formed a company named *Norsk hydro-elektrisk Kvælstofaktieselskab* (Norsk Hydroelectric nitrogen limited, later just Norsk Hydro). The nitric acid plant was located at hydroelectric stations in the towns of Notodden and Rjukan. These plants were profitable until the 1920s; by then, a cheaper

---

[94] William T. Love had begun a canal for hydroelectric power in 1892 in the town of Niagara Fall, but the project was abandoned after about one mile was built. Hydro-power projects had been the backbone of New England industrial development throughout the 1800s.

source of nitrate had been found. And the plants made lots of electricity and had nothing to do with it.

It happened that Wilhelm Ostwald in Estonia (following the earlier work of Charles Kuhlmann) had patented a process for oxidizing ammonia to nitrate with oxygen on a platinum catalyst and looked towards commercializing it circa 1906 using ammonia from coke ovens. But coke oven ammonia was potentially damaging to the catalyst due to impurities. The presence of this technology, led Fritz Haber in Germany to consider the direct conversion of hydrogen (H_2) with nitrogen (N_2) to from ammonia (NH_3).

As a student, Fritz Haber (1868–1934) studied the relative strengths of C-C and C-H bonds on hydrocarbons. In the early 1900s, he and his students began studying the possible reaction of hydrogen with nitrogen:

$$3\ H_2 + N_2 \rightarrow 2\ NH_3 \text{ (very, very slow)}$$

This reaction is thermodynamically favored at ambient temperature, but it is too slow to be useful. At high temperature, ammonia is unstable and the reverse reaction is faster:

$$3\ H_2 + N_2 \leftarrow 2\ NH_3$$

It turns out that in 1884, Henry Louis Le Chatelier had pointed out that when a system is placed under stress, it tends to react to relieve that stress. In this case, application of Le Chatelier's principle was achieved by elevating the pressure of the gases, which would favor forming ammonia (two molecules) over the comparable amount (four molecules) of hydrogen and nitrogen even at elevated temperature. But this idea alone was not enough to make the conversion practical. Thus, in 1909, Haber succeeded where Le Chatelier failed (1904) by developing a catalyst that would work effectively at a relatively low temperature where the equilibrium still favored the product (ammonia). The reaction was scaled-up by Carl Bosch into a successful industrial process in 1913.

The timing was not particularly good because in 1914, German and Austria went to war with Britain, France and Russia. Had the Haber-Bosch process not been available, Germany would have run out of explosives rather early in the war and the death and destruction would not have proceeded nearly as long as it did (1914-1918).[95]

Today, it is estimated that about 50% of the nitrogen in your body has at one time or another gone through a Haber-Bosch plant to make fertilizer. Moreover, it is estimated that about 30% of the people on the planet would starve if this process were stopped (and the survivors would be fighting over natural resources).

---

[95] In 1921, about 200 tons of ammonium nitrate stored at Oppau near Ludwigshafen on the Rein exploded killing hundreds and injuring thousands. In 1947, approximately 2,000 tons of ammonium nitrate (a ship load) exploded at Texas City, Texas killing 581 people.

6.3 The Origins of Petroleum Refining

We have already presented a hypothesis for the accumulation of the liquid pyrolysis products of plants during the formation of coal in Section 5.1. These fluids were trapped in sands under layers of impermeable strata and their low density led to their accumulation in geological dome structures. In this anerobic environment bacteria that acquire energy by oxidizing elemental iron and reducing sulfur dominate:

$$4 \text{ Fe} + S_8 \rightarrow 4 \text{ FeS}_2 \text{ pyrite}$$

Along the way, the bacteria reduce oxygenated molecules (carbohydrates, "CH_2O") that may have accompanied the pyrolysis products producing hydrocarbons, hydrogen sulfide and some hydrogen. (However, much of the hydrogen could be consumed in reducing carbonyl and carboxylic acid and nitrogen compounds.)

$$\text{FeS}_2 + 2 \text{ "}CH_2O\text{"} + 3H_2O \rightarrow \text{FeSO}_4 + 2 \text{ "-}CH_2\text{-"} + H_2S + H_2O$$

Petroleum occasionally is found on the surface of the earth as seeps or tar pits. Some civilizations have used bitumen as a caulking and mortar, others have used the oil as a tonic or medicine, some may have even used it as a lubricant for wheels turning on axels. But in most places, it was just considered as a pollutant of water from springs.

Light after Dark

At some point, probably after the invention and general use of writing (~4000 years ago), the idea of having light after sunset gained more attention. Of course, wood fires and torches were available, but the only convenient source of portable light was the lamp. Tallow and olive oil were introduced in Biblical times and bees wax candles (very expensive) were apparently introduced by 500 BCE.

In the 1700s and early 1800s in the US, whale oil and fish oil became the lamp oils. Then between 1830 and 1850, "paraffin" was distilled from coal and was found to be suitable for making cheap candles in a mechanized process. Chemist James Young (1811-1883) identified a natural oil seep near Riddings colliery (Alfreton, Derbyshire). He began fractional distillation of the seep oil and found he could isolate several products. Soon he discovered that he could obtain paraffin oil by slowly raising the temperature of coal and fractionally distilling. The residue solidified on cooling and was useable as a wax to make candles (patented October 17,1850 in Britain and the US). These patents were upheld and people producing paraffin and related products from coal ended up paying him royalties. Young's Paraffin Light and Mineral Oil Co. Ltd. was active until 1921 employing as many as 4,000 workers. Along the way the paraffin oil lamp was introduced.

Glycerin and Explosives

George Fergusson Wilson (1822–1902) was born into a candle-making company. He patented a process (1842) for isolating fatty acids from tallow by acid hydrolysis and steam distillation. In 1854, he discovered that he could isolate glycerin as a byproduct of this process.

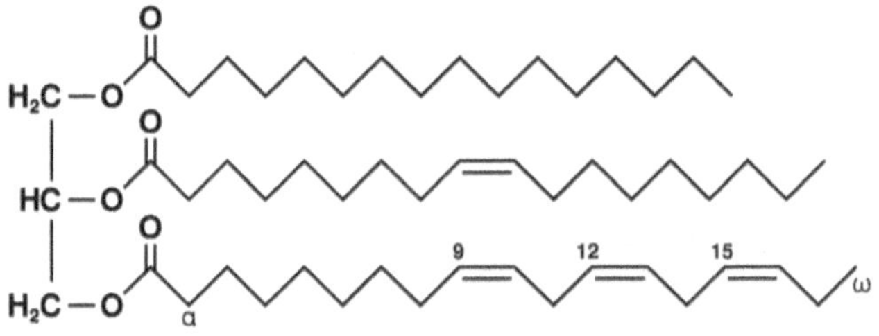

A typical triglyceride with several example fatty acids[96]

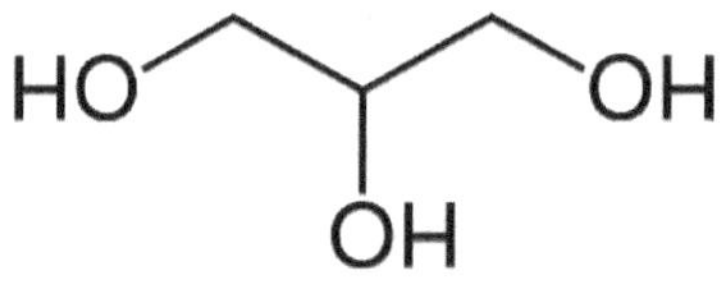

Glycerin (Glycerol)[97]

---

[96]
https://upload.wikimedia.org/wikipedia/commons/b/be/Fat_trigly ceride_shorthand_formula.PNG

[97]
https://upload.wikimedia.org/wikipedia/commons/thumb/d/dc/G lycerin_Skelett.svg/1200px-Glycerin_Skelett.svg.png

Glycerin was used as a carrier for perfume. Being derived in many cases from animal fat, these products encountered some religious prohibitions.

During the 1840s, there was interest in replacing (or supplementing) black powder with other explosives. Théophile-Jules Pelouze at the University of Paris was involved in the study of many things including nitrocellulose. In 1838, Pelouze had dipped cotton balls in concentrated nitric acid and found that the product would explode. In 1846, Christian Friedrich Schonbein (1799-1868) accidentally discovered that a higher level of nitration of cotton (cellulose) could be achieved by using a mixture of sulfuric and nitric acid.

The success of guncotton (nitrocellulose, see below) spurred more investigation among Pelouze's former students Ascanio Sobrero (1812–1888) and Alfred Bernhard Nobel (1833-1896). In particular, Sobrero (then a professor at the University of Turin) produced nitroglycerin (a tri-nitric acid ester of glycerin) in 1847, but he soon found it so unstable that he abandoned work on it.

The Nobel family (who had moved from Stockholm to St. Petersburg) had manufactured munitions for the Russians during the Crimean War (1853-1856), but after the war the family's business neared bankruptcy. Nobel was a student of Pelouze in the early 1850s where he learned explosive technology. Then he spent time in the US during the run-up to the American Civil War (1861-65). He was probably looking for business opportunities for industrial explosives (e.g., tunnels, canals) in the expanding American economy. In 1859, Alfred

Nobel and his parents returned to Sweden, leaving the plant in Russia to Nobel's older brother. In Sweden, Nobel knew that nitroglycerin was a powerful explosive and very easy to make, but its instability limited its utility. Nonetheless, the family proceeded to make and ship quantities of nitroglycerine as an industrial explosive. Unfortunately, Alfred Nobel's younger brother was killed in an explosion of a production plant in 1864.

Nobel apparently came to the idea that if he could find a way to detonate low concentrations of nitroglycerin (which were relatively stable), he could make a safe explosive. To this end, he focused (1863) on inventing a detonator (i.e., blasting cap). Nobel perfected his detonator in 1865. For two more years, Nobel looked for a way to stabilize nitroglycerin so that it would be convenient to handle. He discovered that when it soaked into diatomaceous earth it was stabilized and could be handled very safely.

Nobel soon marketed "dynamite" (1867) and became very wealthy because with dynamite, mining and construction of roads and canals became a much faster process. It also had military applications and (very unfortunately) it became a favorite tool of terrorist. Nobel saw himself as a productive and responsible businessman. However, when one of his brothers died (1888), the obituary he read in the newspaper assumed it was he who had died. More importantly, the obituary was very derogatory painting Nobel as a warmonger and arms merchant. This was not how he wanted to be remembered in history. Not having any heirs, Nobel consulted a pacifist friend (Bertha von

Suttner); and in his will, he established a trust fund and a series of prizes to be awarded in sciences and for peace (1895):

> *"to the person who shall have done the most or the best work for fraternity between the nations and the abolition or reduction of standing armies and the formation and spreading of peace congresses."*

Nitroglycerin of course created a new market for glycerin, which is high-boiling and expensive to separate from the hydrolysis of fats.

Picric Acid and Trinitrotoluene

Picric acid (trinitophenol) was isolates from nitration of various protein materials. It was later manufactured by nitration of phenol. It was discovered to have explosive properties by Hermann Sprengel and became a military explosive in 1871. Trinitrotoluene was actually produced as early as 1863 and was not considered to be an explosive compound. Its potential as a military explosive was recognized in 1891 by Carl Häussermann. This was perhaps inspired with the success shown by Nobel in exploding dynamite (i.e., the idea that a detonator cold be used to set off the explosive charge). Trinitrotoluene and picric acid are shock resistant and were first used to fill artillery shells in 1902. Indeed, TNT is so shock resistant that a delay fuse can be used to cause the charge to explode *after penetrating armor* (e.g., of ships). Bombs can be dropped from high altitudes, accelerate to supersonic speeds and penetrate concrete bunkers without exploding. Thus, TNT has been a mainstay of military explosives throughout the twentieth century.

Petroleum and Kerosene

The history of petroleum utilization, of course, moved at different speeds in different regions. The Chinese distilled petroleum oils in the 500s. In the Middle East, the numerous petroleum seeps and natural wells led to use of this resource primarily as a military weapon.

Modern petroleum refining appears to have been introduced in Romania (Ploesti) in the 1850s. Slightly later, oil seeps in north western Pennsylvania (Titusville) roused the attention of the Pennsylvania Rock Oil Company (George Bissell and Jonathan Eveleth) who were looking for a cheaper source of lamp liquid. They sent Edwin L. Drake to evaluate it and he was motivated to actually drill a well to see if commercially useful quantities of oil were available. Drake hired a salt well driller (salt was obtained from brine found in some aquifers) to drill a well and they began bringing up petroleum in August 1859. By 1862, there was enough oil being produced to warrant a railroad into Titusville. Immediately after the American Civil War (1866), the number of wells increased and the industries needed to supply tools, pipe and pumps both to the oil field and to the oil refineries expanded. The oil boom proved very valuable to the local community.

The oil refineries were crude distilleries where a cut of volatile liquids was captured and some lubricants were provided to the expanding railroad industry. The public's demand for "kerosene" (as the fraction was called) was determined by the lamps and lanterns based on the experience with paraffin oils.

The key elements were a globe at the bottom to hold a supply of kerosene, an adjustable metal roller assembly to move the woven cotton wick up or down and a glass chimney to contain the flame.

Kerosene wicked up from the reservoir as it was burned.

Source: http://www.tgldirect.com/products/unknown/clear-glass-kerosene-oil-lamp-and-chimney-with-eagle-burner-and-wick-17-tall/170400410001/

The lamps and kerosene became very popular. For the first-time average Americans could afford light in their homes for reading and socializing after the sun went down. However, as the business grew, a problem soon appeared. When the lamp was first lit in the evening, it was cool and the most volatile hydrocarbons were considered desirable because they were easy to ignite. But later in the evening, the entire lamp would become quite warm. At this point, possibly when not being observed, the volatile hydrocarbon vapors would escape from the reservoir (independent of the wick, they might even boil and upset the entire lamp). Many fires were thus started causing damage and loss of life in homes. Perhaps the most notorious fire caused by a lamp was the great Chicago, IL fire (October 8, 1871).

Standard Oil

John D. Rockefeller (1839–1937) was raised in near poverty in Cleveland, Ohio. Cleveland became an oil producing center during the American Civil War. Rockefeller formed an oil refinery company in 1862 with partners and bought oil wells in Titusville, PA. Soon the company was very profitable. Rockefeller and some of his competitors joined forces to "stabilize prices" in what was a turbulent market. At the time, there were no laws against monopolies or price fixing. The other thing that Rockefeller did to out maneuver his competition was to standardize his kerosene product so that it did not have large amounts of highly volatile hydrocarbons. By careful distillation and blending, he ensured that his product was consistent and safe to use.[98] To market this consistently safe product, he called the company "Standard Oil" (1870).

---

[98] Today kerosene is characterized by boiling point (between 150 and 275 °C) flash point (between 37 and 65 °C), and autoignition temperature (220 °C). Kerosene for indoor use is preferentially "low sulfur." Chemically kerosene is a mixture of alkanes (straight, branched and rings) 70% with a minor component 25% of aromatics

Along the way, it became apparent that transportation cost was one of the variables that could not be directly controlled. But the railroads had expanded during the War and needed freight customers badly. Thus, Rockefeller was able to secure a long-term low-rate contract with Cornelius Vanderbilt (1794-1877).

This negotiation went in Rockefeller's favor because there was competition in the railroad business between Vanderbilt and Thomas Scott (1623-1881). But in the 1870s, Vanderbilt and Scott, managed to settle their differences and used the same tactic against Rockefeller. Rockefeller was strong enough financially at this point (having acquired assets of most oil refineries in the US) to circumvent the railroads by building pipelines to transport his oil. This created competition to the railroads that Rockefeller could control. From this position, Rockefeller soon displaced the railroad barons and proceeded to become the richest man in the United States.

In 1866, Matthew Ewing and Hiram B. Everest founded an enterprise called Vacuum Oil Company, which used vacuum distillation to remove a mobile, high-boiling fraction of petroleum (useful in lubrication) from the residual tars. Without use of vacuum technology, the high-boiling fraction was just decomposed and fragmented to gases and tar in the distillation process. In 1879, Standard Oil acquired about 75% of Vacuum

---

(which tend to form soot) and very little 5% olefin (which tend to form gums). Note that 37°C is 100°F so that combustible vapors can be achieved at the wick and ignited with a match, but vapors are unlikely to collect outside the chimney.

Oil's assets. The lubricating-oil business (Mobile Oil) became one of the most litigious areas of the petroleum refining industry.

From 1870 to 1890, petroleum refining changed little. The product of interest was kerosene and the heavier oils were mainly used as lubricants in the railroad and steam engine industries. But technology and the economy were about to change. Thomas Edison and George Westinghouse were introducing electric lighting and electric motors and the internal combustion engine was about to find a market in transportation.

The Internal Combustion Engine

Steam engines use fuel (typically coal) to heat water in a boiler and send that steam to drive a piston or turbine through pipes. But early in the steam-driven industrial revolution, various people had seen the possibility of directly using the expanding gases of a combustion process to drive the piston. This technology demanded much more advanced materials and technology before it could be efficiently and economically implemented. As early as 1860, Jean J. Lenoi build a crude engine and attempted to power a wheeled vehicle. But the history of the modern four-stroke Internal combustion engine is complex and controversial.

For generations engineers have called the four-stroke system the Otto Cycle; how that came to be is complicated. Christian Reithmann (1818-1909) was a watchmaker who appears to have been a visionary of automation and robotics. He apparently introduced a small internal combustion engine (bore 9.8 cm and stroke 11.1 cm) into his business before 1860, but was prompted to take out a 1-year patent in 1860 after learning of the work of Lenoi. In 1862, Reithmann filed a patent for a four-stroke engine in Germany, but Alphonse Beau de Rochas (1815-1893) had published the four-stoke system in France in 1861. Importantly, de Rochas had recognized the importance of compressing the fuel-air mixture before ignition. In 1873, Reithmann advanced his designs to something very similar to the modern engine.

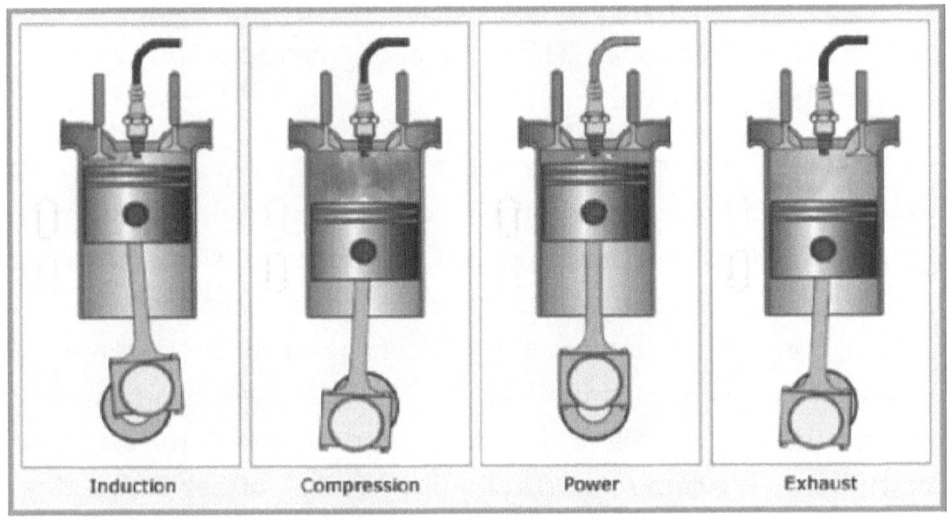

Author: Shaik Moin; Source:
https://shaikmoin.wordpress.com/tag/four-stroke-engine/

Nicolaus Otto (1832-1891) had been working on engine designs since 1862 and introduced his engine in 1875. This was regarded as the *first practical* four-stroke engine design. Thousands of these engines were produced by the company Deutz AG (founded in 1864 and half-owned by Otto) for stationary applications and the engine began to have an impact on industry. Patent litigation soon followed. Otto's patent was overturned in favor of the de Rochas patent in 1886. Reithmann sued Deutz AG and won a settlement. Nonetheless, there was little publicly known about the legal challenges and Otto continued to be recognized as the inventor of the Otto cycle.[99] All of this back fighting was of little concern to Standard Oil. The important point was that after 1875 a new market was being developed for petroleum derived fuels. The original engines actually used hydrocarbon gases, not liquids.

The first device that was intended to evaporate liquids into a stream of air for use in an internal combustion engine was apparently developed by Luigi De Cristoforis in 1876. According to sources, the concept was incorporated in an internal combustion engine developed by Enrico Bernardi to power several small vehicles circa 1882. These examples simply passed air over the surface of liquid hydrocarbons and took the air/fuel mixture into the combustion cylinder. When Gottlieb Daimler (1834–1900) joined Deutz AG in 1876, the company started seriously thinking about engines fueling road vehicles; and for this, liquid fuels were preferred. Indeed, Daimler began designing and building his own engine and was the person that

---

[99] Jutta Siorpaes (2008) *Als die Welt in Bewegung geriet: Christian Reithmann und die Erfindung des Viertaktmotors*

discovered (1877) that Otto's patent was invalidated by de Rochas's parent.

Naturally, Otto and Daimler were not only headed in opposite directions, there was actually personal conflict in the relationship and Daimler was fired in 1880. Daimler had befriended and mentored Wilhelm Maybach (1846-1929) and together they developed a new design for an engine intended to burn naphtha (a more flammable fraction of petroleum distillates than kerosene). At the time, naphtha was marketed as a cleaning solvent and was the fraction of petroleum distillates that had made kerosene dangerous to use before Standard Oil. This engine (small and high RPM) was introduced in 1883-84. It incorporated a float-type carburetor (1885) and used "hot-tube" ignition. An electrical ignition system (by Bosch) was introduced in 1887. The fuel was "heavy naphtha" (bp 90-140°C) that was called ligroin. Light naphtha boils as low as 60°C. The Daimler carburetor patent was soon challenged and lost priority to Edward Butler's (1862–1940) float-type carburetor that had been patented in England in 1880.

Butler and Otto were discouraged from pursuing transportation engines because they realized that they would require use on public roads, which were dominated by horses and posed all sorts of liabilities that were not encountered with stationary engines. Basically, the market for stationary engines was stable and certain, while the idea of the automobile involved much more risk.

Electric Lights and Automobiles

Standard Oil was primarily making asphalt, lubricants (by vacuum distillation), kerosene and naphtha. The real money came from kerosene (lamps lighting) and for practical purposes the other materials were byproducts that were recovered rather than disposed as waste. The advent of the commercial electric lights posed a risk to Standard Oil.

Thomas Edison (1847-1931) had filed his first patent in 1878 and in 1879 found that carbonized bamboo filaments had a reasonable lifetime in incandescent lamps. In 1880 light bulbs were good enough to begin replacing kerosene. But kerosene continued in general use in rural areas into the 1920s (it was expensive to expand the electric grid into sparsely populated areas).

Fortunately for Standard Oil, an engineer at Edison Electric (1891-99) named Henry Ford (1863-1947) became interested in automobiles. While working for Edison, Ford indulged his hobby in building automobiles. Ford left Edison in 1899 and founded the Detroit Automobile Co. But it was underfunded and quickly ran out of capital and closed in 1901. Nonetheless, he built and raced a 26-horsepower car in 1901. It was soon obvious that racing got the public's attention and sold cars. The Ford Motor Company was soon formed (1903), but Ford left his partners and this became Cadillac Automobile Company. Without missing a beat, Ford designed and built at 80-horsepower car that Barney Oldfield drove to victory at the 5-mile Manufacturer's Challenge Cup race in 1902. This brought

Ford national attention. By 1903, Oldfield had broken the 1-minute mile (60 miles per hour) on a closed tract.

Speed was the driver in the automobile world. Carl Fisher championed building a brick-paved course near Indianapolis, IN for the purpose of testing new automobile designs. This track soon became the Indianapolis Motor Speedway and the first 500-mile race was held in 1911. Ford himself benefited from the public's interest in automobiles and he was targeting that market with cars that average families could afford. The legendary Model-T was first sold in 1908. Sales reached 472,000 vehicles in 1916 and the price had dropped to $360/unit.

These cars consumed gasoline and drivers wanted paves roads. Standard Oil had found new markets for its heavy naphtha and asphalt. The company was doing very well...too well. By 1882, the various businesses acquired by Rockefeller were conglomerated into the Standard Oil Trust and dominated every phase of petroleum marketing from the time it was recovered from the ground to the time it was consumed by an end user.

1906 was a watershed year for American businesses. As will be discussed in Part III (below), a number of forces were coming together during the progressive administration of Theodore Roosevelt (1901-1909) to impose government regulation on American business. The food industry had become a national market (you no longer knew the person who raised your food) and people wanted assurances that high standards were maintained. Hence, the 1906 Pure Food Act was passed. Similarly, the wealth and power of Standard Oil was viewed as unfairly setting prices and suppressing innovation, which could only be assured by competition. Thus, in 1906, the US government sued Standard Oil under the Sherman Antitrust Act (of 1890). The complex and intentionally obscure business structure of Standards Oil made this a difficult task, but by 1911, Standard Oil (now headquartered in New Jersey) was broken into 33 different companies.

Two-Stroke Engines

Two-stroke engines (diesel engines) work on a simpler principle. When air is compressed rapidly, the energy put into the compression is conserved by increasing the kinetic energy of the molecules of gas; which means that the temperature of the gas increases dramatically and rapidly. The two-stroke engine dispenses with the separate intake and combustion strokes of a four-stroke engine and depends on a constant flow of air through the intake and exhaust valves while the piston descends to simultaneously sweep in the new charge while sweeping out the exhaust. This requires careful design of the combustion chamber to minimize mixing of intake and exhaust. The down stroke of the piston can be used to pressurize the crank case and force the fuel/air mixture from below the piston into the chamber above the piston.

Harry Ralph Ricardo (1885–1974) was one of the first to explore this arrangement and built a number of small engines 1904-1914. When WWI broke out, he was initially involved in developing engines for British tanks (based on Daimler designs). By 1917, he had a much-improved engine being mass produced. In 1917, he moved into the air ministry and was asked to figure out why airplane engines were sometimes running irregularly.

The Airplane

The dream of controlled powered flight became a reality in the early 1900s. By 1914, reliable airplanes with internal combustion engines of several different configurations (inline, rotary, radial, V) were proving successful. At the beginning of World War I, the military utility of the airplane was not appreciated. They were initially introduced for reconnaissance and opposing pilots were almost collegial toward one another. Then someone remembered that they were the enemy and the race was on. The overall performance (speed and maneuverability) and firepower were rapidly advanced as were the tactics. That is not our interest. What is of interest here is the fact that in the race to produce lighter, more powerful engines, several problems were identified. In particular, the properties of the fuel became critical. The main issue was associated with raising the compression ratio of the engines to improve their efficiency. The compression ratio is the relative pressure of the fuel air mixture when it is ignited (at the beginning of the power stroke) to the normal atmosphere pressure. The following table summarizes the Brake Thermal Efficiency (how much of the energy in the fuel can you convert into useful torque on the propeller at optimum fuel air ratios) as a function of the compression ratio.

Considerations in Design of
Internal Combustion Engines

Brake Thermal Efficiency (full throttle)	Compression Ratio Required	Fuel Octane Required to Achieve Compression Ratio
low	5 to 1	72
25%	6 to 1	81
28%	7 to 1	87
30%	8 to 1	92
32%	9 to 1	96
33%	10 to 1	100

Ricardo began a study of the phenomenon of irregular engine performance at the Air Ministry in 1917. He realized that engine designers had made their engines more efficient during the war by increasing the compression ratio. The compression heating that he was familiar with in two-stroke engines was causing the fuel to ignite prematurely (as the piston was traveling up in the cylinder on the compression stroke) and even explode ('knock") after ignition damaging the piston and cylinder head. Ricardo built a variable compression engine to study the problem.

In the US, General Motors Company was established by William C. Durant in Flint, MI in 1908 to buy automobile businesses more or less the way that Rockefeller bought up oil refineries. In addition to the individual companies, in 1909 GM began a research program (Dayton Engineering Laboratories Company, Delco) to study and develop systems that could be patented and used throughout all of its automobile lines. This activity was consolidated under Charles F. Kettering (1876–1958). In 1916, Thomas Midgley Jr. (1889–1944) came to work for Delco (which was sold) and Midgely ended up in the Research Division of GM (1919).

Concurrent with the work of Ricardo in Britain, the issue of knock was revealed by experimentation at GM. Midgley was given the task of finding a fuel additive that would increase the resistance of gasoline to pre-ignition and knocking. Various compounds were considered. But when the US entered the war

(1917), Midgely looked at making a synthetic high-compression fuel by hydrogenating benzene to make a mixture of benzene[100] and cyclohexane. [101] Kettering realized that GM was outclassed in looking to improve the hydrocarbon gasoline itself and formed alliances with Standard Oil and E. I. du Pont de Nemours.

Research continued at GM. For example, aniline was found to have anti-knock properties; but it had a number of objectionable characteristics as a fuel additive. The team was becoming somewhat desperate and even tried various samples of elements of selenium, tellurium, arsenic, antimony and tin. These compounds all were effective at suppressing knock, but had objectionable properties including strong odors.

It is fairly clear that the GM team (and perhaps no one at the time) had much understanding of flame chemistry. Flames propagate through gas by free radicals initiated by oxygen. Compounds that can scavenge free radicals produced from the fuel (R•) will suppress the combustion (explosion) process:

---

[100] Benzene has very good antiknock properties, but freezes at a temperature that make it impractical for airplane engines. Benzene also is a good solvent and damages rubber (e.g., fuel lines, float valves in carburetors and self-sealing liners for fuel tanks).

[101] Dietmar Seyferth. 2003.The Rise and Fall of Tetraethyllead. 2. *Organometallics*. 22:5154-5178.
https://pubs.acs.org/doi/pdf/10.1021/om030621b

$$R\bullet + R\text{-}M \rightarrow R\text{-}R + M\bullet$$

Where the alkyl radical (R•) is able to extract H• or R• from a hydrocarbon but M• is not reactive enough to do that.

In December 1921, the team tested tetraethyl lead and found that (aside from acute and chronic toxicity with no warning properties) it was compatible with fuels and was exceptionally effective against knock. In 1922, GM hired du Pont to manufacture tetraethyllead (from sodium lead alloy and ethyl bromide).

However, a long series of tests by GM revealed that lead oxides formed in the engine with a variety of adverse effects. Thus, trials were run to find a way to prevent this accumulation of lead oxide. It was determined that dibromoethane effectively prevented the accumulation of lead in the engine, because under the conditions of combustion $PbBr_2$ was volatile and would pass into the exhaust system and then out with the exhaust gases. On the economic side, obtaining bromine from brine wells in Michigan (operated by Dow Chemical) was considered undesirable and GM and Standard Oil (ESSO) formed Ethyl Corporation (1923) and worked out a process for chlorinating sea water to obtain bromine to make dibromoethane. The bromine production plant was built at Kure Beach, NC. By this point it was realized that ethene (ethylene) was among the gases recovered from petroleum refining. GM obtained a "use" patent for tetraethyl lead as an anti-knock additive and Standard Oil (ESSO) obtained the patent for dibromoethane.

It was not long before ethyl chloride replaced ethyl bromide in the manufacture of TEL. Ethyl hired du Pont to operate the manufacturing plants. These plants were brought up to speed much faster than desired because of the economic benefit that was anticipated. Unreasonable exposure of some Du Pont workers to tetraethyl lead resulted in death and illness. Nonetheless, "Ethyl gasoline" went on sale in early 1923 in Dayton, OH. The "Ethyl Fluid" package added to gasoline was:

TEL   63.30%
Ethylene Dibromide 25.74%
Ethylene Dichloride 8.73%
Dye[102] and impurity 2.23%

Because of concerns about toxicity, in 1924, TEL addition to fuel was limited to about 3.0 mL/gallon. When TEL was first introduced in 1923-25 its price was about $3.00 *per* pound. This dropped rapidly to about $0.75 *per* pound by 1935 and stabilized at about $0.50 *per* pound during the war years (1940-45).

---

[102] The dye was included to warn users that the product contained toxic TEL. Blue dye was specified in 100-octane U.S. Army grade "Fighting 2-92" and U.S. Navy grade "M 302."

The Inter-War Years 1919-1939

One of the more obscure figures of this period was Frederick B. Rentschler (1887-1956). He was a Princeton graduate (1909), who during WWI, had entered the Army and ended up overseeing production of aircraft engines. After WWI, the Wrights founded the Wright Aeronautical Corporation and Rentschler became its president. The Curtiss Aeroplane and Motor Company of Buffalo, New York founded by Glen Curtiss, was the main competitor of the Wright company. Curtiss had focused on high powered airplanes that were developed immediately after WWI using liquid-cooled engines. The premier "muscle" engine of the day was the Curtiss D-12 (V-1150), a 12-cylinder V-design engine with 1150 cubic-inch displacement and 435 horsepower. This engine was used to power the PW-8 pursuit airplane purchased by the Army in 1923.

http://www.airwar.ru/enc/fww1/pw8.html

An even larger 500 hp V-1400 water-cooled engine was available from Curtiss by 1925. Air-cooled engines of the day were generally only about 100 horsepower. The Wright company was falling behind in the horsepower competition.

Rentschler wanted to follow the radial engine design but was unhappy with the support of the Wright Aeronautical Company along these lines and resigned from the company in the summer of 1924. Rentschler took his ideas and talent to the Pratt & Whitney Tool Company in the spring of 1925. There he developed the first large radial, the 9-cylinder Pratt & Whitney Wasp (R-1340), which was completed on 24 December 1925. The engine produced 425 horsepower weighing only 650 pounds, and was impressive enough for the Navy to contract for 200 engines in 1926. The Pratt & Whitney company then moved ahead with the 525 horsepower Hornet engine that weighed only 750 pounds (1.42 pounds *per* horsepower). Although the radial design might appear to be counterproductive in streamlined airplanes, the radial design eliminated the cooling system/fluids and radiator required with the inline designs. These coolant systems increase weight and militance and reduce reliability and created combat vulnerability.

In 1926, Curtiss formed a joint venture with Wright (Curtiss-Wright) in part to export obsolete U.S. airplanes to countries in South America and to develop engines that would compete with Pratt & Whitney.

Rentschler's departure was a blow to Wright, but they fought back, and on 21 May 1927, Charles Lindbergh completed his solo flight across the Atlantic in the *Spirit of Saint Louis* powered by the 223 hp Wright Whirlwind J-5C radial.

Curtiss-Wright hired a well-known racing pilot named Jimmy Doolittle, on leave from the Army, to demonstrate their obsolete P-1s to South American countries. He proved to be quite a successful salesman. While flying in Mexico selling airplanes for Curtiss-Wright, Jimmy Doolittle had an unpleasant experience with tetra-ethyl lead. He carried a container of tetra-ethyl lead in his airplane to enhance the octane of local fuels in Latin America. The container broke and Doolittle was almost overcome by the toxic fumes.

Supercharging

Up until 1930, the horsepower race had depended upon making bigger and bigger cylinders with more and more displacement. But the General Electric Company and others had been working on ways to increase the amount of fuel-air burned in the same cylinder volume as a way of increasing the horsepower of the engine without increasing its displacement. The concept was called supercharging and it had the added benefit of compensating for the reduction in atmospheric pressure that airplanes experienced as they increased their altitude. As a matter of fact, that was why most work prior to 1930 had been done on supercharging, i.e., to compensate for lost horsepower with altitude rather than to increase sea-level horsepower. The reason supercharging initially focused on compensation for altitude was that the available fuel was only of about 60 octane and the engines would "knock" if they were over-pressured (e.g., supercharged).[103]

Straight Run Gasoline

The original refining process utilized until about 1930 was nothing more than fractional distillation. Out of the research with variable-compression test engines, it was determined that different pure compounds responded differently to compression. Thus, an octane isomer (2,2,4-trimethylpentane) was arbitrarily assigned the rating of 100 and n-heptane was given the rating of 0. Most pure hydrocarbons fell within this range. The early internal combustion engines had low compression ratios and

---

[103] The dream of coast-to-coast commercial flight service required that airplanes be able to fly over the Rocky Mountains at altitudes of about 20,000 feet. It was realized in WWI that engine performance dropped off with altitude and General Electric engineers, led by Dr. Stanford N. Moss, set out to solve this problem with a supercharger. In a famous series of tests in 1917 and 1918, they dragged a 350-hp Liberty airplane engine to the top of Pike's Peak (14,109 feet) to conduct test with a turbosupercharger. *Turbo*supercharger refers to a compressor driven by the exhaust gases of the engine rather than by mechanical connections (such as gears).

used very low octane fuel (e.g., the naphtha fraction of petroleum has an octane of around 50).

Boiling Range (degrees Celsius)	% Recovered by Atmospheric Distillation	Chemical Composition
0-30	2%	C1-C4
30-200	15-30%	C5-C12
200-300	5-20%	C12-C15
300-400	10-40%	C15-C25
over 400	10-50%	over C25

Through the 1920s, straight run[104] gasoline with addition of TEL was regarded as the cure-all for engine knock. Depending upon

---

[104] The modifier "straight run" was added later to distinguish gasoline obtained from the continuous distillation run from gasoline obtained by blending or chemical modification. Since WWII, gasoline has been *synthesized* from petroleum rather than *refined* from it. The petroleum companies have retained the term "refining" in part to disguise their relationship to the rest of the synthetic chemical industry especially

the composition of the hydrocarbons in the fuel, the allowed 3.0 mL of TEL per gallon would raise the octane rating about 10 to 20 octane numbers. The fuels with the lowest octane numbers are most improved by TEL such that the best hydrocarbon gasoline might only be enhanced about 10 octane numbers. However, when the supercharged engine designs went over an effective compression ratio of about 8 to 1 (the mechanical displacement compression ratios were about 6.5 to 1), adding 3 mL TEL *per* gallon of gasoline would not raise the octane of straight run gasoline (octane about 55-60) adequately.

In 1925, the U.S. Army Air Service decided to test the turbosupercharger on a 510-hp Packard 1A-1500 engine. This power plant combination was put into the XP-4 biplane and tested on 27 July 1927. The supercharger worked, but the XP-4 did not perform as well as the un-supercharged version (ceiling 22,000 feet). Why? Well, the supercharger weighed 800 pounds, which constituted 30% of the airplane's weight. Nonetheless, work continued and on 12 April 1930 Captain Hugh Elmendorf led the 95th Pursuit Squadron of the Army Air Corps to a world record for formation flying of 30,000 feet, which far surpassed the previous record of 17,000 feet. In the early 1930s, Trans World Airlines also spent a substantial amount of time developing and refining high altitude technology.

----

once the environmental movement began. Even as late as 1998, the U.S. Environmental Protection Agency labored under the assumption that gasoline is a natural fraction of petroleum.

Because compressing the air with the supercharger caused adiabatic heating, the air was cooled in a radiator called an intercooler before being directed to the carburetor.

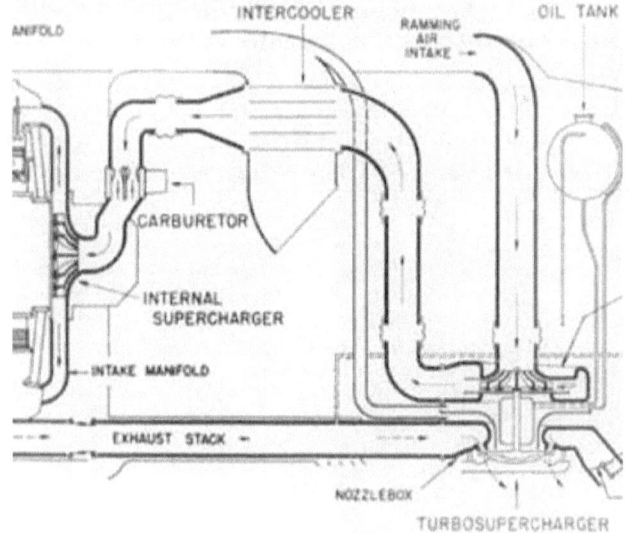

http://rwebs.net/avhistory/opsman/geturbo/geturbo.htm

Thus, by the early-1930s, TEL was no longer the cure-all additive for gasoline. To have higher octane, there would have to be a basic improvement in the refining process. It was clear, that petroleum refiners needed to improve the octane quality of their basic hydrocarbon product. Stopgap measures such a isolating the "heart cut" of low boiling isopentane (octane 90) from the straight run fuel were tried, but this component was only about 15% of natural petroleum, and once it was removed, the remaining naphtha had a pitifully low octane of about 30, which made it almost useless even as a motor fuel. Thus, high-octane (aviation) fuels became a specialty item about 1930 because, supercharging technology was pushing beyond the limits of ordinary straight-run gasoline supplemented with TEL.

Increase in Engine Performance (Compression Ratio) Reflected in Typical Octane Ratings

Year	U.S. Army Air Corps	British Air Ministry	Civil Aviation
1922		51	
1925		55	
1928		58	
1930	71		73
1931	78	68	73
1932	82		73
1933	87		79
1934	90	69	80
1935	91		83
1936	93	70	85
1937	95		87
1938	98	87	89

Cracking, Reforming and The Houdry Process (1930-)

In the United States, petroleum was readily available and American research into higher octane gasoline was different from the work in Europe. The European developments after 1930 will be discussed after summarizing the American situation. The first major step, in making the petroleum refinery more than just a distillery, was taken by Eugene Houdry who was born in Domont, France and moved to the U.S. in 1930. With the main interest of obtaining more liquid fuel from petroleum, he took the residue from petroleum boiling above 370°C at atmospheric pressure and vacuum distilled it. The liquid was not volatile enough to be used as fuel in ordinary engines, but he discovered that under these conditions, he could "crack" the large molecules into lower molecular weight molecules. This process also happened to increase the olefin content of the fuel.

$$C_xH_{2x+2} \rightarrow C_aH_{2a+2} + C_bH_{2b} \text{ where } x = a + b \text{ } \textit{Cracking}$$

It was also recognized that cracking increased gasoline's octane. During this same period (1933-1939), some "reforming" of straight-chain aliphatics into branched-chain aliphatics was accomplished by rearrangement reactions that accompanied thermal cracking.

Houdry improved the un-catalyzed (*i.e.*, thermal) cracking process by finding catalysts that facilitated the molecular changes. Using the Houdry catalytic cracking process, fuel of about 90-octane could be obtained after addition of TEL. There was, however, a major problem with cracked fuels. The olefins in the cracked fuel tend to polymerize forming gum. The gum collected in the carburetor, which was a much more serious matter for airplanes than for automobiles. Thus, while cracked hydrocarbons became popular as component **of** *"motor fuels,"* they were usually specifically prohibited from *"aviation fuels."* In some cases, the cracked fuels could be hydrogenated to eliminate the olefins, but this also lowered the octane number.

Low molecular weight (C2 -C4) hydrocarbons were a byproduct of the Houdry cracking process. Initially, these gases were just another waste, but work was soon underway to find uses for them. The two most important components of this mixture were butane and butene. Normal (n-) butane consist of four carbon atoms in a chain with two hydrogen atoms on each of the two center carbons and three hydrogen atoms on the two end carbons (C-C-C-C). n-Butene has the four carbons in a chain, but two hydrogen atoms have been removed and a double bond in found either between the first and second carbon (1-butene, C=C-C-C) or between the second and third carbon (2-butene, C-C=C-C). In 2-butene, there can be a subtle isomerism relating to the spatial orientation of the two end carbons relative to the double bond.

Butane also exists in an isomeric (iso-) form where there is a central carbon and the other three carbon atoms are attached to it. The central carbon only has one hydrogen atom and the other three carbons have three hydrogen atoms each. There is only one iso-butene isomer. Two carbons share a double bond and one of these has two hydrogen atoms while the other carbon has two methyl groups ($-CH_3$)

6.4 Petroleum Fuel Leading into World War II

Shell Oil Company and Jimmy Doolittle

In 1929, Shell Oil Company was eager to develop a high-profit product to sell to the military as low-octane automobile fuel consumption dropped with the Wall Street stock market crash. Automobile fuel was about 50-octane and generally sold for a few pennies a gallon in 1934; while aviation fuel was 91-octane and sold $0.15 *per* gallon. Shell recruited Jimmy Doolittle to represent their product, high-octane aviation gasoline for use with high-compression and supercharged engines. In the Shell fuel, the octane rating was increased by synthesizing iso-octane (as described below) rather than by adding tetra-ethyl lead. There was only one catch, high-octane fuel was difficult and expensive to manufacture.

But Doolittle saw the potential of high-octane fuel and was effective in winning over the military.[105] Doolittle wanted to set the octane-rating standard as high as possible for aviation fuels.

---

[105] In addition to being a great flyer, Doolittle also had a Ph.D. in aeronautical engineering from MIT.

He and Shell settled on 100-octane as a round figure. Doolittle was convinced that he could eventually get the Air Corps to standardize its engine specifications on this figure. In 1933, Shell built a million-dollar plant in Wood River, Illinois to turn out 100-octane fuel. In 1934, 100-octane aviation gasoline cost the Air Corps $2.40 *per* gallon. In spite of the obvious difficulties of a cash-strapped Air Corps trying to increase the cost of its fuel by more than a factor of 10, Doolittle and Shell persisted. Shell invested in three more plants in 1934. The 100-octane fuel was achieved in part by engineering the hydrocarbons themselves and by adding tetra-ethyl lead as an anti-knock agent.

In March 1935, the results of experiments with 100-octane fuel at Wright Field with the Boeing P-26 were published in the *Journal of Aeronautical Sciences* under the title "Aircraft Engine Performance with 100-Octane Fuel." In November 1936, the Army finally convened an expert panel under Col. J. T. McNarney to examine the issue of standardizing military octane requirements. The committee included representatives of engine makers Wright, Pratt & Whitney and United Aircraft; and fuel suppliers Shell Oil, Phillips Petroleum and Standard Oil. The committee resolved "*that 100-octane fuel would be as the standard for the Air Corps by January 1, 1938.*" Standard Oil won the first delivery bid for 100-octane gasoline, but Shell Oil soon received its share of business. Curtiss-Wright announced that they were going to design and build an engine that would make the most of 100-octane fuel. In fact, what they did was to adjust the effective compression ratio (supercharging) and timing of the R-1820 Cyclone to optimize its performance with 100-octane gasoline.

Polymerization of Butenes to Manufacture Octane (1934-1941)

From the fractional distillations and particularly from the cracking processes (thermal and catalytic), oil refineries produced a large quantity of butane (C_4H_{10}) and butene (C_4H_8). These volatile gases were originally either "flared" (*i.e.*, burned at the end of an outlet pipe) as waste or burned in the refinery for fuel. About 1930 (even before Houdry's work), the Universal Oil Products company began experimenting with these butenes especially with an eye towards polymerizing them to iso-octene, which can be hydrogenated to iso-octane.

$$2\ C_4H_8 \rightarrow C_8H_{16}\ \textit{Polymerization}$$

$$C_8H_{16} + H_2 \rightarrow C_8H_{18}\ \textit{Hydrogenation}$$

There are several processes for conducting the polymerization. The original process used high temperatures, but catalysts were found to make the reaction more efficient. The catalysts were all acids, but different acids and conditions have different advantages. The bottom line is yield and economics. Several pilot plants were built using a thermal polymerization process for turning butene into octene between 1931 and 1934. Soon, a unified process for manufacturing octane emerged:

    (1) petroleum was used to make butene;

    (2) the butene was polymerized to iso-octene; and

    (3) the polymers were hydrogenated to make isooctane.

By 1934, encouraged by Jimmy Doolittle's arguments that the Air Corps could be convinced to move to a 100-octane aviation gasoline standard, Shell Oil Company (subsidiary of Royal Dutch Shell) built a petroleum-to-octane plant at their East Chicago Refinery. This plant used the Universal Oil Products Company "cold acid" catalyzed polymerization that worked selectively for isobutylene yielding high-purity iso-octene, which was hydrogenated to 100-octane iso-octane. Shell Oil quickly introduced plants for using "hot acid," which polymerized both iso- and n-octene. It was upon this process that Shell hoped to supply the needs of the Army Air Corps; but when the Army Air Corps accepted 100-octane as their standard in 1938, this new market initiated an octane competition between the major oil companies. The search for a better way to make octane was begun and Shell would soon find its processes obsolete.

Before it was superseded, the Shell Oil process spread through the industry in the U.S. and overseas. It was still the method of choice in Britain when the war started in 1939 and through the Battle of Britain in 1940. However, in 1941, Imperial Chemical Industries (ICI) of England applied the Houdry dehydrogenation process to butane to make more butene.

C_4H_{10} butane → C_4H_8 butene + H_2 *Dehydrogenation*

This allowed them to obtain more octane from a barrel of oil. The hydrogen produced by dehydrogenation was used to hydrogenate the octene polymer.

Also, in 1939, Houdry, a patriotic Frenchman who was wounded in WWI, returned to France and helped the French government adapt his process to the production of high-octane aviation gasoline. Unfortunately, this was not the best time to introduce high technology into the French petroleum industry as they were soon overrun by the Germans in 1940. Houdry returned to America where he helped develop the process for making butadiene for synthetic rubber. By 1942, about 90% of the aviation fuel in the world (Axis and Allies) was manufactured from butene/butane obtained by cracking petroleum in the Houdry process. Dehydrogenation of butane to yield additional butene was used in the U.S. and Europe into 1941-42, but it was soon made obsolete.

6.5 Seeds of Conflict and Strategy Europe (1930-1939)

In the 1930s, the situation is Europe was quite different. The United States had an essentially unlimited supply of petroleum and the British could obtain petroleum fairly readily from the western hemisphere or Middle East (assuming they could keep control of the oil fields in the Persian Gulf and the lines of communication could be kept open). But mainland Europe was starved for petroleum feedstock. The only two countries in Europe that had substantial petroleum reserves were Rumania and Poland. Germany, France, Italy, Austria, Hungary and Czechoslovakia had/have minimal petroleum reserves and were turning to alternate fuels especially ethanol and coal-derived liquid fuels. In 1935, Germany imported 1.2 million tons of petroleum fuel and Italy imported about a fourth that much. France imported about 1 million tons of fuel in 1935.

In Germany, domestic petroleum provided only about 5% of its peacetime needs in 1935. The Germans were also concerned about a short supply of potatoes for ethanol fermentation stock. However, within a few hundred miles, the Polish army was mounted on horses and carried lances to protect what was described in 1935 as a "plentiful supply of Polish crude."[106] Rumania was the second largest petroleum producer in Europe. In 1935, Rumania's distilling capacity was 11.8 million tons *per* year and their cracking capacity was 1.5 million tons *per* year; their actual 1935 production was 8.8 million tons. It was known that the octane rating of the straight run gasoline from Poland and Rumania was low (60-61 octane).

In Britain, aviation gasoline was typically a good straight run fraction fortified with tetraethyl lead. This combination yielded

---

[106] Gardner, F.H. and Hardiman, E.W. 1938. Motor Fuels in European Countries in *The Science of Petroleum* by A.E. Dunstan, A. W. Nash, B.T. Brooks, and H. Tizard Volume IV. Oxford University Press, London. pp. 2407-2413.

octanes up to about 77 to 87, which were the British and German standards. The British did very little cracking until 1938. However, Britain had large stocks of coal and during the process of making coke, a "light oil" is distilled off that contains benzene, toluene, ethylbenzene and xylenes (BTEX).

Coal (CH) → carbon (coke) + light oil (BTEX)

The Europeans called this *benzole* (in Britain) or *benzol* (in Germany). The British also tended to call gasoline "benzine." The British used their coal-derived *benzole*, which has an octane of about 130, as a blending stock. They would routinely use blends of 10% to 20% *benzole* with straight run gasoline for aviation fuel. One of the problems with going to higher levels of *benzole* was freezing of the benzene and naphthalene (a trace impurity) in the carburetor. In 1938, the French also set 70 to 87 octane as their standards for aviation gasolines.

Interestingly, high-octane-rated methanol and ethanol were used as the fuels when the British and French set out with their high-compression racing airplanes to establish speed records. But in addition to limited supply, and high cost, the alcohols had high fuel consumption (limiting range) and alcohol also caused corrosion problems in the fuel systems.

In Germany, synthetic liquid fuel plants were under construction in 1935 (e.g., Fischer-Tropsch, see below) and the Nazi government was predicting that it would be self-sufficient in petroleum within a few years. But by this time, it was established that the octane rating of most of this synthetic gasoline (obtained indirectly from coal) would only be about 47, even lower than straight run distillates. Moreover, it contained enough olefins to cause a "gum" problem; and *if the synthetic fuels were hydrogenated to eliminate the gum problem, the octane of the paraffin was only 12.* So, the Germans could produce motor fuel for ground vehicles, but aviation fuel was more difficult to produce. The Germans were very interested in cracking technology and were happy to acquire Houdry's process after they overran France (1940).

But the demand for high-octane was an artifact of the Otto-cycle (4-stroke) internal combustion engine. The Germans worked hard at using diesel (2-stroke) engines for airplanes. Diesel engines were useful in heavy equipment such as trucks and required development of fuel injector systems (instead of carburetors). Fuel injector technology showed up in some of the German aircraft engines. It was especially useful for water-methanol injection, which actually was used in large enough quantities by the Germans to be considered a substitute fuel (not just an octane-enhancing additive).

To escape the "octane trap," the Germans also showed interest in the jet engine at an early stage. In Britain, Frank Whittle (1907–1996) patented a design for a centrifugal-compressor jet engine in 1932; and in Germany, Hans von Ohain was inspired to make a more advances design in 1935. While national pride may go to the British, the Germans were quick to recognize the importance of this technology and had jet airplanes flying in August 1939 (He 178). Nazi Germany invaded Poland in September 1939, which forced Britain and France to declare war on Germany.

Synthesis Gas

In the 1800s, it was discovered that coal could be burned in the presence of steam to produce a gas containing mainly carbon monoxide (CO), hydrogen (H_2) and methane (CH_4) with some carbon dioxide. This mixture is flammable and was produced as "water gas" or "town gas" for use in street lamps and in other roles that natural gas (methane) now plays. This gas also fueled the first internal combustion engines and is the reason that the fuel for internal combustion engines is called "gas" or "gasoline."

Uses for this gas were pursued and it soon became obvious that it could be used as an important raw material for chemical synthesis. Thus, it acquired the name "synthesis gas" inside the chemical industry. The basic reactions were as follows:

Coal (CH) + H_2O → CO (38%) + H_2 (52%) + CH_4 (10%) + CO_2

Of course, the reaction occurs only at high temperature with the aid of catalysts. It was found that more hydrogen could be obtained by another high temperature catalyzed reaction (the water shift reaction):

$$CO + H_2O → CO_2 + H_2$$

Whereas the typical water gas had a ratio of CO to hydrogen of about 1.0 to 1.3; many of the subsequent synthetic uses worked better with ratios carbon monoxide to hydrogen ratios of 1.0 to 2.0. So, the water shift reaction was used to enrich the hydrogen content of synthesis gas. (In practice, sometimes hydrogen could be obtained more economically by cracking coke-oven gas.) With these reactions, German industry could start with coal and water and readily produce hydrogen and carbon monoxide.

Through 1913, the only conversion of synthesis gas to hydrocarbons that had been achieved was the conversion of synthesis gas to methane,

$$CO + 3\ H_2 \rightarrow CH_4 + H_2O$$

but in that year, a patent claimed to convert synthesis gas to liquid hydrocarbons.

The Fischer-Tropsch Process

Two chemists who had experimented with synthesis gas, F. Fischer (1877-1948) and H. Tropsch (1889-1935), tried unsuccessfully to verify the reaction under the claimed conditions of the 1913 patent. Nonetheless, they found that they could obtain liquid hydrocarbons by passing synthesis gas over an iron-alkali catalyst at 400°C and several atmospheres of pressure. They called the product "*synthol*" and it contained various oxygen-containing aliphatic compounds. The *Badische* Company took this line of development to make methanol.

$$CO + 2\ H_2 \rightarrow CH_3\text{-}OH$$

It took Fischer and Tropsch more than a decade to discover and develop a catalyst that would convert synthesis gas to petroleum-like hydrocarbons:

$$8\ CO + 17\ H_2 \rightarrow C_8H_{18}\ (paraffins) + 8\ H_2O$$

By the 1926-28 time period, they were using a catalyst of cobalt and copper, which converted synthesis gas to petroleum-like hydrocarbons at about 300°C. They were able to obtain about 130 grams of liquid hydrocarbons (called *"kogasin"*) *per* standard cubic meter of synthesis gas. The typical products were 4% gases (*e.g.*, propane and butane); 62% gasoline-like hydrocarbons; 23% diesel-like hydrocarbons, and 11% waxes. The octane number of the raw gasoline fraction was 47 and it contained about 25% olefins. When this mixture was hydrogenated to eliminate the gum-forming olefins, the octane rating went down to 12 because it contained almost exclusively straight-chain (normal) aliphatic hydrocarbons.

The next step was to take this process out of the laboratory and develop it as a commercial process. This was done mainly by the *Ruhrchemie A.G.* Company who mastered the chemical engineering problems.

Between 1927 and 1932, the Fischer team developed the catalyst and was working to replace the batch process used in the laboratory with a continuous process for commercial manufacture. In the continuous process, the synthesis gas was swept past the catalyst as a moving gas stream and a certain amount of hydrocarbon was formed in each pass through the reaction vessel. The conversion was measurable by the contraction in the volume of the gas as liquid was produced. The engineers tried to obtain as much conversion of the synthesis gas to hydrocarbon on each pass as possible, but total yield could be raised by sending the gas through several stages of reactors (multiple passes). The work continued in 1932-1934 using a Ni-MnO-Al₂O₃ catalysts on *kieselguhr* (diatomaceous earth used as a support). This catalyst produced about 70 grams of *kogasin per* pass at 200°C and 1 atmosphere of pressure. While the chemists worked to improve the catalyst, the engineers were busy solving the heat transfer problem. Namely, the formation of hydrocarbon and water from carbon monoxide and hydrogen is very exothermic. The catalyst rapidly heated up and either melted the reactor or charred the hydrocarbon to coke.

Ultimately, two reaction vessels became popular in Germany depending whether the reaction was run at atmospheric pressure or at moderate pressures. In the low-pressure system, each reactor unit was about 5 m long x 2.5 m wide x 1.5 m tall. As many as 50 to 100 of these units might be in a single reactor building. Each reactor unit was fabricated of 555 ordinary steel plates (1.6 mm thick by 2.5 m x 1.5 m) separated by 7.4 mm. The plates were penetrated by 630 parallel steel pipes (34 mm OD and 29 mm ID). The system worked like a radiator core. The catalyst was placed in the open space between the plates and cooling water was run through the pipes. The synthesis gas was passed through the catalysts like air through a radiator. The heat was generated by the reaction and carried away by the cooling water. Because the entire reaction was run at about 200°C (well above the boiling point of water) and was constantly producing heat, the water was pressurized and carried to a boiler where the steam could be used in the processing plant.

The synthesis gas contracted as it passed through the catalyst bed and was conducted to another reactor unit for further reaction. Along the way, the lost volume of gas was made-up with fresh synthesis gas (carbon monoxide to hydrogen ratio about 1.0 to

1.5). Typically, two or three stages of reaction were used. At the end of the process, the mixture exiting from the last stage was mainly hydrocarbon vapor. Waxes tended to build up on the catalyst over a period of weeks and these were removed by periodically shutting down the system and passing liquid hydrocarbon through the catalyst bed to dissolve the wax. With care, the catalyst could be kept active for months.

The reactor for the moderate-pressure synthesis used a different design. The catalyst was placed in concentric steel pipes and the synthesis gas flowed into the center pipe (about 4 m long) and out the annulus between the two pipes (the outer pipe was sealed at the bottom). These units were ganged together in groups of about 2,000 and immersed in cooling water in a drum 4.5 m tall and 2.7 m in diameter. The higher-pressure units produced more wax, which was used as a cracking stock and for lubricants.

The chemists continued to develop the catalyst; and between 1933 and 1939, the catalyst that was developed was composed of cobalt-thorium oxide-magnesium oxide on *kieselguhr* (100:5:8:200 by weight). All the German plants between 1938 and 1944 operated using this catalyst and the *Ruhrchemie A.G.* process at 180 to 200°C and either 1 atmosphere (low-pressure) or 10 atmospheres (moderate-pressure). The typical yield was 150 grams of hydrocarbon *per* standard cubic meter synthesis gas (carbon monoxide to hydrogen ratio 1.0 to 2.0) and the 60 to 100 volumes of feed gas were sent through 1 volume of the catalyst *per* hour. The catalyst lasted up to 150 to 200 days.

Of course, nothing is ever quite that simple. Coal contains sulfur and the synthesis gas contained traces of organic and inorganic sulfur compounds that would poison the catalysts. Thus, the raw synthesis gas had to be purified by removing sulfur compounds before it could be used. The sulfur removal process was itself a work of chemical and engineering art.

In summary, a synthetic gasoline plant as operated in Germany during World War II included: A conventional water gas generator followed by a compressor. The compressed water gas passed through units to remove hydrogen sulfide and organic sulfides. Part of the water gas was converted into synthesis gas by the water shift reaction. The hydrogen-enriched gas was sent to the first stage of Fisher-Tropsch reactors. After the first stage, high boiling liquids were removed in a condenser and fresh water gas was introduced to make up the pressure. The process continued through one or two additional stages of reactors. The final product was sent to a cooler condenser and activated carbon where most volatile organics were removed. Non-condensable gas (hydrogen and methane) was flared or burned for heat in the plant. The product yields were as follows:

Gasoline	Diesel	Wax
Low-Pressure Process (1 atm)		
56%	33%	11%
Moderate-Pressure Process (10 atm)		
35%	35%	30%

The plants were sometimes located near *benzol* plants (discussed below) and the *benzol* plants were able to supplement the hydrogen needs of the process in lieu of the water shift unit. During the war (1939-45), the limitations on synthesis gas production (*e.g.*, hydrogen availability) was actually the limiting factor in the rate of these plants.   It typically took several days to start up a *synfuel* plant as the reactors were brought up to temperature and the continuous process was balanced.

Fuel Directly from Coal (The Bergius Process)

It was known that coal (with and empirical formula of essentially CH) produced "light oils" consisting almost exclusively of aromatic hydrocarbons (benzene, toluene, ethylbenzene and xylenes; BTEX) when heated without air. These oils were a byproduct of coke (C) formation, which was a key ingredient for conversion of iron to steel.

Coal (CH) → carbon (coke) + light oils (BTEX)

If this reaction is attempted in the presence of oxygen, polynuclear aromatics (PNA) and phenols are obtained (coal-tar creosote):

Coal (CH) + O_2 → naphthalene, PNA + phenols + H2O

While coal tar creosote is a valuable product and the naphthalene that is produced is a unique route to an important class of compounds, naphthalene and the phenolic mixture do not constitute an attractive liquid fuel. It was discovered that if coal is heated with hydrogen and catalysts, the yield of aromatic hydrocarbons is increased and the yield of coke was suppressed.

Coal (CH) + H_2 → BTEX

The hydrogenation of coal is more rapid and efficient if the coal is finely divided and suspended in a hydrocarbon liquid. The mixture of aromatics produced by this process (*i.e.*, the Bergius process) was called *benzol* (German for benzene) and is an excellent high-octane fuel-blending-stock (*i.e.*, octane about 130). Unfortunately, the yield of oil *per* ton of coal is low and the process is energy intensive. Nonetheless, manufacture of *benzol* became a key part of the German strategic fuel economy because of its octane value. The biggest problem was that the process required hydrogen and the only large sources of hydrogen were from synthesis gas and cracking. The requirements for hydrogen for the Bergius process taxed the available synthesis gas formation systems.

I. G. Farbenindustrie Company engineered the Bergius process. Powdered (less than 80 mesh) German brown coal was mixed with about 10% by weight red mud from bauxite purification (iron oxide/caustic) and dried to about 5% moisture content. The solid mixture was crushed and mixed with an equal volume of heavy oils to form a paste. This paste was pumped with about 400 cubic meters of hydrogen *per* ton of paste through a pre-heater at high pressure (300 atmospheres) into the reactor, which was maintained at about 480°C.

The *Farben* engineers designed the reaction vessel as a cylinder 1.2 m (5 feet) in diameter and 18 m (60 feet) tall. The pressure retaining wall was protected from the heat by a layer of insulation. Heat in the reactor was provided by the exothermic hydrogenation reaction. The hydrogenation reaction was controlled by controlling the rate of hydrogen introduced. The typical rate was 1000 cubic meters of hydrogen *per* ton of coal. The reaction mixture passed continuously through three or four successive reactors. The through-put was about 0.45 tons of ash *per* cubic meter of reactor volume *per* hour. The reaction mixture emerged from the last reactor still under high pressure and it was dumped into a separation chamber (catch-pot). From here, two oil streams emerged: (1) As the hot mixture cooled, the vapors were removed from the hot catch-pot to a cool catch-pot and condensed. These vapors were the fuel product. (2) The heavy oil slurry remaining in the hot catch-pot was allowed to cool and was discharged as a liquid at atmospheric pressure. The heavy oil was mixed with lighter oils collected as vapors from the catch-pot and centrifuged. The solids from the centrifuge were coked in a rotating kiln to recover more of the heavy oil. All of the heavy oils obtained were combined and used to make new coal-catalyst paste for the process.

The product stream from the first stage of coal hydrogenation accounted for about half of the weight of the material reaching the separation chamber (*i.e.*, the yield of volatile liquids was about 50% of the weight of coal). After removing the tar acids and bases by washing with aqueous acids and bases, the neutral oil (crude *benzol*) was typically about 60% aromatics (*benzol*), 30% paraffins, and 10% olefins.

The crude *benzol* from the first stage was catalytically hydro-refined in a second stage of production consisting of two steps:

> In the first step, the liquid was saturated (*i.e.*, the olefins were converted to paraffins, but the aromatics were not affected) by passing over a tungsten-nickel sulfide catalyst supported on activated alumina. The reaction was carried out at about 300 atmospheres and 400 °C, which was

maintained by addition of hydrogen. The through-put was about 1 ton *per* cubic meter of catalyst *per* hour.

In the second step, the aliphatic hydrocarbons were dehydrogenated and cracked using similar conditions with a dehydrogenation catalyst that was tungsten sulfide on a clay support.

Overall, most of the cyclo-paraffins and cyclo-olefins in the crude *benzol* obtained after stage one, were dehydrogenated to aromatics and the non-cyclic hydrocarbons were cracked to C1 to C4 hydrocarbons that were taken off as a gas. Thus, the finished *benzol* product from the second stage was a highly aromatic mixture. The finished *benzol* could be blended to enhance the octane of Fischer-Tropsch fuel. Moreover, toluene could be isolated from the mixture and used as the starting material for trinitrotoluene (TNT).

Alkylation of Isobutane (1935-1949)

Back in the US, Standard Oil of New Jersey had carefully followed the efforts of Jimmy Doolittle and Shell Oil to get the Army Air Corps to adopt the 100-octane standard in 1938. They knew the basic polymerization technology used by Shell (*circa* 1934) and they knew that if they could find a more economical way to accomplish the same transitions from cracked petroleum to octane *via* butane/butene, they would be able to undercut Shell's prices. The method they developed between 1935 and 1941 was called "alkylation."

Alkylation is the direct reaction of isobutane with olefins (especially n-butene and iso-butene) to form octane:

C_4H_{10} **butanes** $\rightarrow$ C_4H_{10} **isobutane**   *Isomerization*

C_4H_{10} **isobutane** + C_4H_8 **butene** $\rightarrow$ C_8H_{18} *Alkylation*

The reaction was similar to the polymerization reaction and was catalyzed by acids, but it eliminated several steps in the reaction sequence (*i.e.*, hydrogenation). Thus, it was much more efficient and economical than the Shell polymerization process.

Chemistry, Business and Politics of Alkylation

Everyone must remember that up until December 1941, the United States was officially (and legally) neutral in the war between Britain and Germany. Well into the 1930s, trade was encouraged with the countries that became the Axis. It was official U.S. policy to ship gasoline to Japan (into the summer of 1941) almost to the time of the attack on Pearl Harbor (7 Dec. 1941). Nonetheless, after the war started and ran its course, various politicians and what have become known as the "revisionist historians" forget this point. Starting with the Truman[107] Committee on Military Affairs hearing in 1942, the major oil companies in the U.S. (*e.g.*, Standard Oil of New Jersey) were persecuted and publicly chastised for trading and

---

[107] US Senator Harry S. Truman (D) from Missouri.

cooperating with I.G. Farben. Antony C. Sutton's 1976 book *Wall Street and the Rise of Hitler* is typical of this view.[108] Joseph Borkin (1978)[109] gives a much more balanced discussion of I.G. Farben and Standard Oil in *The Crime and Punishment of I.G. Farben*.

However, in response to the Truman Committee on Military Affairs' accusations in 1942 alleging Standard Oil's conduct was "treasonous," a Dr. R. T. Haslam (a director of Standard Oil of New Jersey) published an article in *The Petroleum Times* (25 December 1943) entitled "Secrets Turned into Mighty War Weapons Through I.G. Farben Agreement." The spin of the article was that the U.S. had gained important information from the Germans (I.G. Farben) through the various business agreements and working relationships. The article may have exaggerated the point and because some of the work (such as ongoing alkylation research) was classified as "secret," the Haslam article avoided mentioning it altogether.

---

[108] Sutton Antony C. 1976. *Wall Street and the Rise of Hitler*. '76 Press, P.O. Box 2686, Seal Beach, California 90740. 220 pp. It should be noted that Sutton and other "revisionists" have as their basic premise that there is an international conspiracy of intellectuals and wealthy people to somehow control the world; and they interpret all events in that context. I will not address that premise here.

[109] Borkin, Joseph. 1978. *The Crime and Punishment of I.G. Farben*. The Free Press a Division of Macmillan Publishing Co. New York, NY. 250 pp.

Understandably, when the Haslam article found its way into Germany (which was, in 1944, losing the war and was being plastered by American bombers), officials at I.G. Farben were worried about having to explain their alleged duplicity to the Gestapo (whose reputation was even more feared than the Truman Committee). Naturally, I.G. Farben officials formulated their own propaganda to convince their critics (the Gestapo) that I.G. Farben had actually out-foxed the Americans. (See Borkin, 1978, pp. Chapter 4, *The Marriage of I.G. and Standard Oil under Hitler*, pp. 76-94.) Sutton (1976) published a lengthy excerpt from a memo dated 6 June 1944 (the day of the Normandy invasion) by a Mr. von Knieriem an official of I.G. Farben (Sutton, 1976, pp. 70-72). The first couple of paragraphs ramble about how the Americans invented the octane concept (*circa* 1923) and how this helped the Germans. By 1939, any oil chemist or engine designer would have had to be incompetent to not know this history. The von Knieriem memo does contain several interesting points, however, (quoted from Sutton, 1976, pp. 71-72):

> *"Shortly before the war, a new method for the production of isooctane was found in America — alkylation with isomerization as a preliminary step. This process, which Mr. Haslam does not mention at all* [as noted above, the research was now classified in the U.S.], *originates in fact entirely with the Americans and has become known to us in detail in its separate stages through our agreements with them, and is being used very extensively by us."*

As von Knieriem states, the information *"...has become known to us...in its separate stages..."* Basically, I.G. Farben was able to take routinely available information and piece together what to do and how to do it.

The revisionists historian's thesis that Standard Oil gave secrets to I.G. Farben falls down very quickly and totally if the contemporary (1935-1940) technical press is consulted. Probably the most relevant single document was published in the *Journal of the Institution of Petroleum Technologists* (volume 24, pp. 303-325) in 1938. This article by S.F. Birch and coworkers of the Anglo-Iranian Oil Co. was entitled *"Saturated high-octane fuels without hydrogenation. The addition of olefines to isoparafins in the presence of sulfuric acid."* You cannot be much more specific than that. This paper starts with a review of the published literature, which includes reference to the 1936 publication in the *Journal of the America Chemical Society* by Ipatiev and von Grosse. An entire section is devoted to the reaction of isobutane with isobutene describing reaction conditions, yields and octane numbers obtained. The paper ends with a discussion of how to best formulate 100-octane fuels. ***This was in the open literature of 1938.*** The Germans undoubtedly knew about alkylation as it is mentioned in two contemporary German technical publications (Mader, 1942 and Jantsch, 1941 and 1943). [110] And, they cite open

---

[110] Marder, Maximilian. 1942. *Motorkraftstoffe* (motor fuels). Springer-Verlag, Berlin. This document was published in the U.S. in 1945 under the Alien Property Custodian Act by J.W. Edwards of Ann Arbor, Michigan. There is a section entitled *Isoparaffinische Benzine durch*

literature information from the late 1930s for the key insight into the reaction mechanism (*i.e.*, Birch and Dunstan 1939. *Transactions of the Faraday Society* vol. 35, page 1013; a British chemistry journal).

As discussed above, the synthetic hydrocarbons produced by the Germans during the war were largely <u>straight chain</u> aliphatics. According to Weil and Lane (1948, pp. 112-115)[111]:

> *"The C3-C4 fraction* [of synthetic hydrocarbons from synthesis gas] *generally constitutes about 13 per cent of the crude liquid product obtained from the normal-pressure synthesis and about 7 per cent of that produced in medium-*

Addition von Olefinen an Paraffine (iso-paraffin gasoline from addition of olefins to paraffins) pp. 403-418. There is an entire chapter on *Hochleistungskraftstoffe* (high-test fuels) including a discussion of additives to improve octane number. pp. 459-518.

Jantsch, Franz. 1941, 1943. Kraftstoff-Handbuch (Fuel Handbook in German). Published by Franckh's Verlagshanding, Stuttgart, Germany. Franz Jantsch was the Technical Testing Expert for Oppau at Ludwigshafen am Rhein. The first edition was published in 1941 (forward signed November 1940) and the second edition was published in September 1943.

[111] Weil, B.H. and Lane J.C. 1948. *Synthetic Petroleum from the Synthine Process.* Remsen Press Division, Chemical Publishing Co., Inc. New York, NY.

pressure operations. This fraction is of special interest in any consideration of the Synthine process because of the prospects it offers for the manufacture of high-octane polymer or alkylate gasoline…."

"It may be of interest to review here some of the uses to which the Germans have put the C3-C4's produced in their commercial processes. In some of the commercial plants, the C3-C4 cut was absorbed in sulfuric acid and hydrolyzed to alcohols, some incidental polymer being also formed and used for blending in motor fuel…."

"The Germans do not appear to have put very strong emphasis on the production of polymer or alkylate from Synthine C3-C4's, although they were somewhat active along this line. Evidently the strong demand for C3-C4 alcohols and the fact that coal hydrogenation was strongly favored (politically) for gasoline production both served to de-emphasize this development. However, research on the 'Iso-Synthesis' [to make high-octane branched-chain aliphatics from synthesis gas] *was continued throughout the war, and polymer plants had been built. The Castrop-Rauxel plant of Klocknerwerke A.G., for example, was found to possess a polymer gasoline unit capable of processing 25 tons of C3-C4 fraction per day at about 2950 pounds per square inch by the phosphoric acid process. On the basis of 10 days of experimental runs, a 45 percent yield of polymer gasoline was expected."*

Overall, polymerization and alkylation appear to have been minor contributors to the German octane pool by the end of the war and most of that must have been based upon the C4 - hydrocarbons obtained from refining natural petroleum. *But German chemical plants available to be inspected after the war in 1946 were not necessarily representative of what was in progress in 1942-43.*

Whatever information I.G. Farben received through 1941 was rapidly becoming obsolete as research was shared within the U.S. extensively during the war and new technologies emerged. Many patent applications were submitted to the U.S. Patent Office and there were all classified as "secret" until 1946. At that time, they were declassified and most of them were published in a single 1948 compendium by the American Chemical Society (Egloff and Hulla, 1948) with this introduction:

> *"A number of oil companies pooled their scientific and technical knowledge on alkylation during World War II, in order to obtain maximum alkylate production from each refinery. This information was secret during the war. It has since been declassified…"*

The original technology utilized sulfuric acid as the catalyst, but this soon gave way to hydrofluoric acid in the US.

By 1946, there were 32 alkylation plants operating in the US using sulfuric acid as the catalysts, 27 plants were using hydrofluoric acid catalyst, and one plant was using aluminum chloride. The product of these plants was called "alkylate" and was used as a blending stock. It was about 24% 2,2,4-trimethylpentane. Other isomers (2,3,4- and 2,3,3-trimethylpentanes) added another 25%. By 1945, the U.S. capacity to produce alkylate was 178,000 barrels *per* day.

The standard 100-octane fuel used by the U.S. Army Air Corps during the war consisted of approximately 50% high-octane straight run petroleum fraction blended with isooctane "alkylate" with 3.0 mL tetraethyl-lead *per* gallon.

Chemistry, Business and Politics of TEL

Of course, TEL technology was widely known in the 1930s by the Japanese and Germans. However, Mr. von Knieriem of I.G. Fgarben (in his memo of 6 June 1944 that was intended to protect Farben from the investigations by the Gestapo) goes on to state the following:

> *"Above all,* [Germany benefited from] *improvements of fuels through the addition of tetraethyl-lead and the manufacture of this product. It need not be especially mentioned that without tetraethyl-lead the present methods of warfare would be impossible. The fact that since the beginning of the war we could produce tetraethyl-lead is entirely due to the circumstances that shortly before, the Americans had presented us with the production plans, complete with their know-how. It was, moreover, the first time that the Americans decided to give a license on this process in a foreign country (besides communication of unprotected secrets) and this only on our*

*urgent request to Standard Oil to fulfill our wish.
Contractually we could not demand it, and we found out later
that the War Department in Washington gave its permission
only after long deliberation."*

Sutton (1976) uses this statement to imply the treason of
Standard Oil, but, in von Knieriem's own words, *"…the War
Department in Washington gave its permission…"* Standard Oil was
not violating the law or secretly collaborating with the Nazis. In
fact, Sutton (1976) discusses this incident at length. Contrary to
von Knieriem's statement, it did not happen *"shortly before* [the
war];" the process began in 1934 shortly after Hitler came to
power, but before the rearmament of Germany was obvious. It
occurred through the Ethyl Gasoline Corporation (formed in
1924 by Standard Oil and General Motors Corporation to
manufacture and sell tetraethyl-lead). Because of the work of
Jimmy Doolittle, who both tried to sell high-octane gasoline to
the Germans for Shell Oil and argued that the Americans should
adopt the 100-octane standard for military aviation (probably
assuming that the Germans would soon do so). The Army Air
Corps was more sensitive to the transfer of technology to the
Germans in 1934 than was the Commerce Department. Thus,
while the U.S. Government allowed the transfer of technology, a
minority (but an informed minority) opinion was expressed by
the Army Air Corps on 15 December 1934 to the presidents of
Ethyl Gasoline and General Motors (quoted from Sutton, 1976,
pp. 189-190):

"I learned through our Organic Chemicals Division today (15 December 1934) that the Ethyl Gasoline Corporation has in mind forming a German company with the I.G. to manufacture Ethyl lead in that country.

"I have just had two weeks in Washington, no inconsiderable part of which was devoted to criticizing the interchange with foreign companies of chemical knowledge which might have military value. ...

"It should seem, on the face of it, that the quantity of Ethyl lead used for commercial purposes in Germany would be too small to go after. It has been claimed that Germany is secretly arming. Ethyl lead would doubtless be a valuable aid to military aeroplanes.

"I am writing you this to say that in my opinion under no conditions should you or the Board of Directors of the Ethyl Gasoline Corporation disclose any secrets or 'know how' in connection with the manufacture of tetraethyl lead to Germany.

"I am informed that you will be advised through the Dyestuffs Division of the necessity of disclosing the information which you have received from Germany to appropriate War Department officials."

Although this letter is perceptive, so was General Billy Mitchell, who was courts marshalled (1925) for his claims that Japan would eventually attack the U.S. by bombing Pearl Harbor. At the time that the Air Corps letter was written (1934), the best the Air Corps could say was "It has been claimed that Germany is secretly arming." So, if the U.S. Government is not willing to take a position on the rearmament of Germany, why should a private corporation be expected to act as if the U.S. were at war? Hindsight is often "20/20." Nonetheless, on 12 January 1935, E. W. Webb (the President of Ethyl Gasoline Corporation) responded to the Army Air Corps by offering to insert a clause in the contract to guard against technology transfer. Ultimately, Ethyl G.m.b.H. was formed to manufacture tetraethyl-lead in Germany and a firm was also formed in Italy. Nonetheless, Germany was still not self-sufficient in tetraethyl-lead, as Germany imported 500 tons of TEL from the U.S. in 1938. Although this may sound like a lot, the U.S. production in 1941 was 50,000 tons and in 1944 the U.S. produced 100,000 tons of TEL. Moreover, the limiting component of the anti-knock liquid (at least in the U.S. and the U.K.) tended to be the ethylene dibromide (not TEL). There was not adequate supply of bromine from brine wells in most countries; and to meet US demands, most bromine was being obtained from seawater processed in North Carolina.

Further Wartime Developments

There is nothing magic about isooctane (2,2,4-trimentylpentane). It is only one of many highly branched isoalkanes with a molecular weight (volatility) convenient for use as internal combustion fuel. As a matter of fact, among the isoalkanes, isooctane has one of the lower octane-numbers and is one of the least susceptible to octane enhancement with tetraethyl lead. It was recognized as early as 1926 that 2,2,3-trimenhylbutane (a.k.a., triptane) had a higher unleaded octane than isooctane (*i.e.*, 100 at lean mixture and 165 at rich mixture). Through 1938, no more than gallon-size quantities of triptane had been synthesized for research. In 1941, the Army tested triptane in the Pratt & Whitney R-1830 engine and obtained about 1,400 hp (Schlaifer and Heron 1950)[112] as compared with 1,200 hp with 91/96 octane grade. Moreover, with 3 mL of TEL *per* gallon, the octane was increased by about 45 octane numbers.

At that time, the synthesis of triptane (by Dow Chemical) was *via* an organo-magnesium alkylation, which would have consumed vast quantities of magnesium metal (2 pounds of magnesium metal *per* gallon of fuel) and cost about $30/gal. The General Motors Company conducted further research and built a triptane plant that could produce about 150 gallons *per* day in 1943 without using magnesium. This plant provided testing samples until 1945. According to Schaifer and Heron (1950, p. 657), "*...very few laboratory engines used for fuel evaluation with supercharging were strong enough to permit triptane + 4 cc lead to be appraised.*" That is, they did not have compression ratios high enough to fully test the fuel. Thus, tests were done with blends of triptane and standard 100 (lean)/ 130 (rich) octane fuel. Such a blend produced 2,800 hp in an Allison V-1710 engine. Without TEL, triptane has little advantage over isooctane as a blending stock.

---

[112] Schlaifer, R. and Heron, S.D. 1950. *Development of Aircraft Engines and Development of Aviation Fuels (Two Studies of Relations between Government and Business)*. Harvard University, Boston, MA.

With introduction of the turbojet engine in 1945, the era of cost-is-no-object high-octane aviation fuels was over and triptane was never developed commercially.

Platinum Reforming

Almost as soon as the war was over (April 1949), Universal Oil Products Co. announced the "platforming" process in which a platinum catalyst converted low-octane linear aliphatic naphtha into high-octane branched aliphatic and naphthene (i.e., ring compounds, like cyclohexane) alkanes. The cycloalkanes can be dehydrogenated to benzene, toluene, ethylbenzene and xylenes. These products were branded as "platformate" and became the basis of eliminating TEL from commercial use in the 1980s.

6.6 Petroleum and the Conduct of World War II

World War II (1939-45) was an amazing episode in technological development. It had to be, the existence of what we will call "Western Civilization" and democracy in general was at stake. The Nazis Germans and the Imperial Japanese made no secret that their intensions *to subjugate and rule submissive countries* was both unlimited and ruthless. Ironically, the Nazis under Hitler were face-to-face with the USSR under Stalin. There was a brief period when it appeared as though these two powers might work together[113]; but in the end there can be only one supreme

---

[113] Importantly, the Molotov–Ribbentrop Pact (August 23, 1939) between Nazi Germany and the USSR was not merely a non-aggression pact. That document contained secret clauses that facilitated a mutual occupation of Poland. As the German invasion

dictator and because of their proximity, Hitler's Nazi Germany and Stalin's USSR clashed early in the war. Nazi Germany invaded the Soviet Union (22 June 1941). Had the Nazis and Imperial Japanese prevailed, they would have eventually clashed as well, but being on opposite sides of the world meant this could not happen until all other nations were conquered and that never happened.

Thus, I have gone into detail above and will go into detail below that you probably do not expect to find in a chemistry book. *Remember this book is about the influence of chemistry on civilization.*

Petroleum in the European Theater

As mentioned above, Germany had substantial coal reserves, but no significant access to petroleum. The initial moves of Hitler were directed at Austria and Czechoslovakia. These were easily explained political actions that enlarged the German sphere with

---

unfolded (September 1939), the USSR took up positions in about 50% of eastern Poland on the pretext they were protecting the Poles. The Soviets never attacked the Nazis. Subsequently, many prominent Polish citizens were shipped to Siberia and most Polish army officers were interned and soon murdered by the Soviets in the Katyn forest (May 1940). As the Nazis fought the Soviets in 1942-45, it became desirable for the Nazis to reveal the duplicity of the Soviets to their Anglo-American Allies. Thus, on April 13, 1943 the Nazis announced that they have recovered the bodies of 4,443 Polish officers in mass graves with evidence that they been summarily executed by the Soviets. Ironically, if the secret parts of the Molotov–Ribbentrop Pact had been known to Britain and France, they would have been compelled to declare war on the Soviet Union as well as Nazi Germany in 1939. The political implications of that are mindboggling.

little external complaint. But they also opened a pathway towards substantial petroleum reserves in Romania, Ukraine and the Baku Region between the Black Sea and the Caspian Sea down to the Persian Gulf. Unbeknown to the British and French, Germany and the USSR had secretly agreed (Nazi–Soviet Pact of August 23, 1939) to partition Poland. Poland's oil was in the extreme south-eastern region and fell under the Soviets, but Germany was already committed to acquire petroleum from Russia.

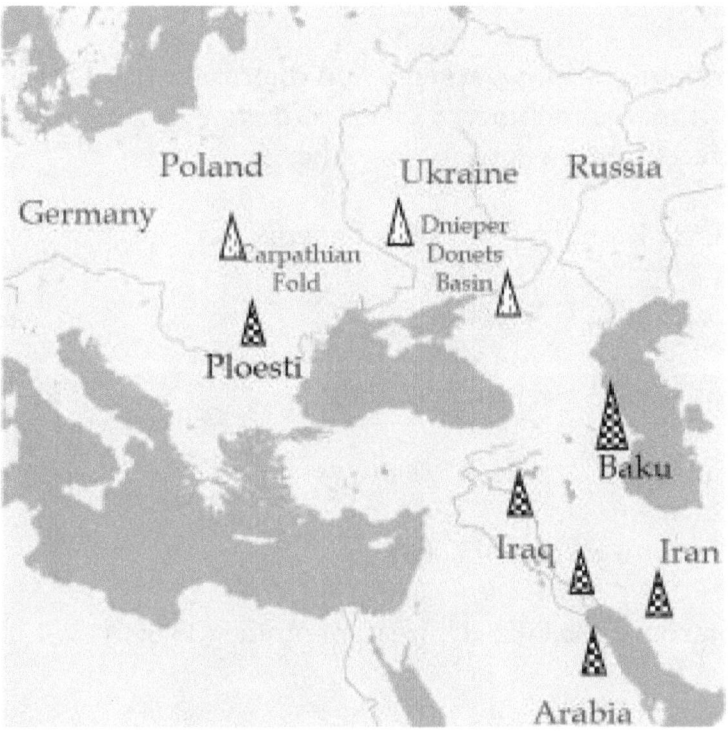

Clearly, Hitler intended to deal with the western allies (France and Britain) first and then secure Romanian oil before turning on his partner Stalin. The time table was uncertain. The conquest of

France in the spring of 1940 was much quicker than could have been expected, but Britain (unexpectedly) escaped and refused to give in. With Germany's failure to achieve air superiority over the English Channel in September of 1940, Hitler was forced to accept a persistent (but contained) Britain in the west and immediately turned east to secure Romania (October 8, 1940). Capturing Ploesti was an essential step if he hoped to invade the USSR.

Ploesti (1939- March 1944)

The petroleum refining, storage and distribution facilities at Ploesti, Rumania included 8 major refineries (potential monthly production in metric tons in parenthesis):

Astra Romana (146,000) - Phoenix (65,000)

Concordia (110,000)

Romana Americana (92,000)

Colombia Aquila (45,000)

Greditul Minier (45,000)

Stardard (36,000)- Unirea Sperantza (33,000)

Xenia (22,000)

Dacia Romana (15,000)

About 15 miles northwest of Ploesti, the Steaua Romana refinery added another 125,000 metric tons *per* month refining capacity.

Thus, the region had a total potential (*e.g.*, design) capacity to refine 734,000 metric tons *per* month. The central Ploesti refineries were located around an area of about 19 square-miles, which included large marshaling yards for shipping and receiving by train and pumping stations. The large Astra Romana facility was centrally located and most plants pumped their production to Astra Romana where it was put onto trains or pumped by pipeline to the Giurgiu terminal on the Danube River. From Giurgiu, the refined petroleum was transported up the Danube by barge into central Europe.

The Italian Distraction

The British had reluctantly dealt with the French fleet (Mers-El-Kebir on 3 July 1940) before it could fall into Nazi hands. Britain also controlled Egypt and the Suez Canal. The Italians had vanquished the obsolete army of Ethiopia and Mussolini appears to have assumed that Britain could easily be dislodged from Egypt bringing him the prize of the Middle Eastern Oil fields; and thus, elevating his status as an ally to Hitler. Concurrent with the final efforts by Germany to dominate Britain at home, Italian troops invaded Egypt from Libya (September 13, 1940). They failed.

The British, indeed, drove the Italians back; and in March 1941, the Germans were forced to get involved in North Africa. Concurrently, April 1941, the Germans pushed the British out of

Crete and Greece to ensure the safety of the Ploesti facilities. Here Hitler became distracted by the North Africa option. Hitler's logic is not clear; but perhaps with his recent successes, he saw the Italian plan as a way to simultaneously weaken Britain, and flank the Soviet Union with a quick strike into the Persian Gulf. Germany attempted to break through to the Persian Gulf until the British won a substantial battle at El Alamein (23 October – 11 November 1942).

In reality, all the activities in the Mediterranean area pale in comparison to the titanic invasion of the Soviet Union begun on 22 June 1941. Considering the uncontrollable (and perhaps unexpected) entry of the United States into the European war, after being attacked by the Japanese in the Pacific, the efforts in North Africa became a diversion of effort that may have made Hitler's broad offensive against Russia unwinnable. But, Hitler's ego and fear of a resurgent and motivated Russia drove him to expend enormous effort on largely symbolic campaigns (Leningrad, Moscow and Stalingrad). None of these political objectives were as strategically important as Baku, which was never attacked because of the battle at Stalingrad. By this time (1942), the US was in the war and Germany's entire southern plan became a fighting retreat back into Tunisia.

The Americans launched two heroic (though futile) bombing attacks on Ploesti from North Africa in 1942 and 1943. The first mission flown against the petroleum complex at Ploesti did little damage. But, the 1 August 1943 attack by unescorted B-24s from

Libya had damaged Astra Romana (power house and cracking plant) temporarily reducing its capacity about 50%, The Phoenix Orion plant had been hard hit as were the Greditul Minier, Colombia Acquila and Steaua Romana plants. However, damage from this single raid was quickly repaired. Because of downstream reserves of fuel in storage and the fact that the plants seldom worked at design capacity, the single attack in August 1943 had *no strategic effect on fuel supplies*. Only continuous attacks would have strategic effects. The single attack, of course, prompted repair work, but it also set the Germans to building defensive structures and planning for a serious defense.

To harden the refineries against bombing attacks, the Germans built blast walls between the process units. The blast walls were made of brick and started at the ground about 6 feet thick tapering to about 2 feet thick as they rose to heights of 20 feet or more. The blast and fragments of a bomb were thus limited to damage of a single unit. The fires could also be contained. Perhaps more importantly, spare parts and repair crews were located in the area so that the systems could be brought back on line when they were shut down. The target was also hardened by moving in many heavy flak guns (88-mm, 105-mm and 128-mm) and preparing airfields to station *Luftwaffe* interceptors in the area if/when serious air attacks began.

By the spring of 1944, the *crude oil refining capacity* of the repaired and fortified Ploesti facilities was back to the range of over 700 metric tons *per* month. The actual production involved converting 458,000 metric tons of crude oil (about 60 % of capacity) to 177,000 metric tons of gasoline each month.

U.S. Strategic Doctrine

I have written a complete multi-volume analysis of the American strategic bombing campaign in the Mediterranean theater (*Memoirs of Foggia*, current edition on line at Amazon/Kindle). Here we only need the highlights.

The ability to strike the industrial capacity of the enemy was a major new twist introduced by bombardment from the air. The US (more than other countries) focused on targeting specific industries, products and plants. To prioritize potential targets for bombing, a group of economists were employed and the decision-making went in three directions: (i) German aircraft production, (ii) transportation (rails and roads), (iii) oil production. While attacking airplane manufacturing was initially the focus (1943), these targets (e.g., ball bearing plants) were actually difficult to hit and easy to move, hide or harden. Up until the invasion of German-occupied France (June 6, 1944) the advocates of bombing transportation were in control. The basis for this decision is beyond the scope here.[114] After the

---

[114] Transportation was critical to tactical defense of Europe. Attacking railroads had an important spin-off: The Nazis were expected to use chemical weapons to defend against invasion. If transportation (train traffic) is being attacked, it was very dangerous for the Germans to

invasion was successful, (summer 1944), with the German's no longer a real threat to reach Baku or the Persian Gulf and US bombers in Foggia, Italy, the time was finally right to eliminate the Nazi oil supply.

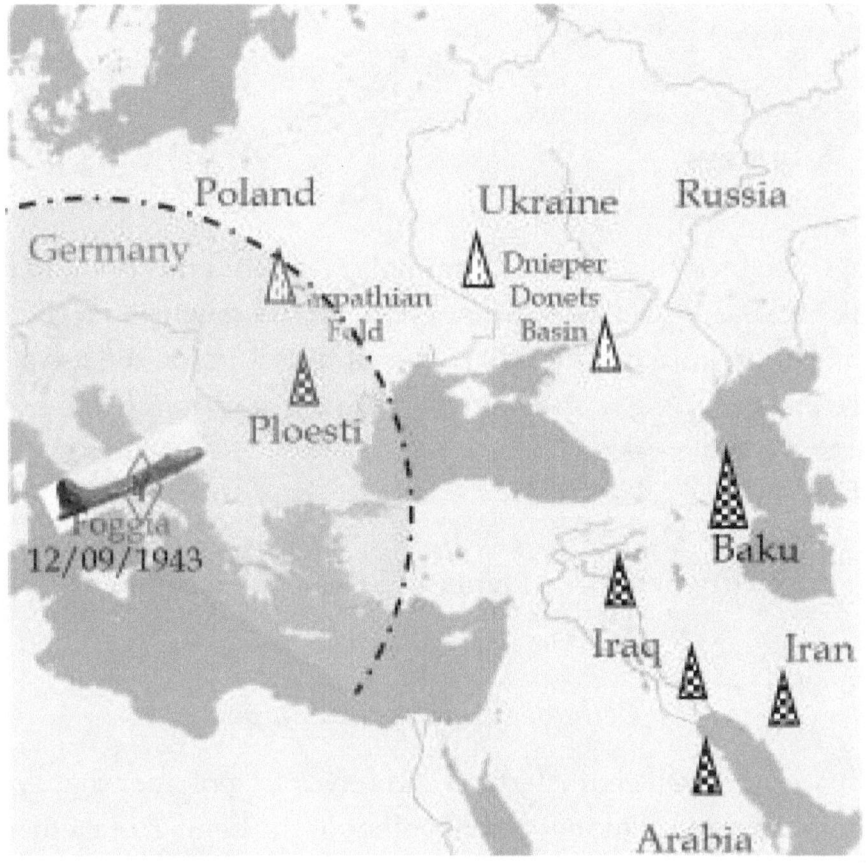

move chemical weapons through its own or occupied territory. A rail car full of mustard gas broken open by allied bombing would be devastating to the German efforts. The allies did not know it, but the Germans had developed nerve agents based on the work of Gerhard Schrader (1936).

When the Americans and British reached Foggia, Italy in December 1943 with heavy bomber that could reach Ploesti, Romania *on a daily basis*, the war was effectively won.

Checkmate/ Game Over.

7.0 Polymers

I have declined to discuss natural polymers until this point in the book because we actually did not understand much about their composition until the late 1800s. Up until this point, the use of these materials (e.g., cotton, wool, silk, etc.) was dominated by trial and error necessity.

7.1 Polymers from Plants and Related Compounds

Cellulose and Wooden Ships

Cellulose is a well characterized carbohydrate polymer that differs from starch in that arrangement of the hemi-acetal ether linkages between glucose units (R-O-R) are on opposite sides of the 6-membered ring allowing a long linear chain (e.g., 10,000 glucose units). In contrast starch, forms a spiraling polymer because the connecting oxygens are on the same side of the ring. Glycogen is a branched starch used to store glucose in the liver.

Author: NEUROtiker; Source: Wikimedia Commons

The chains of cellulose polymer are held together by hydrogen bond.

You may wonder how wooden ships with curved hulls were made. The answer is that when steam is pushed into the wood, the water separates and relieves tension in the cellulose strands making wood very plastic. Boards can be bent into graceful curves. When the excess water is lost and the temperature lowered enough to re-establish hydrogen bonds between cellulose strands, the strands are locked into a new shape.

https://www.culturenordic.com/products-page/Vikings

King Cotton

Cotton as almost pure strands of cellulose is produced by a family of plants. It appears that these plants use carbohydrate as cotton (in lieu of sugary or starchy fruits) to attract birds. The cotton seeds are closely intermingled with the fibers. The birds can use the fiber for nesting and coincidentally disperse the plant seeds. In Volume II of my book *Antebellum* (2019) I have gone into the history and industrialization of cotton processing in great detail. Here I only offer a brief outline.

Cotton was originally a product of India and was moved by Arab traders to Egypt. The Muslim conquest brought cotton to Spain (750 CE) and the reconquest (1099 CE) carried cotton and the knowledge of its spinning and weaving across Europe. Some varieties of cotton were already present in the new world, but cotton is prone to hybridization and new varieties have been introduced over time. Egyptian cotton is noted for its long fibers which were relatively easy to spin dominated the early production into the 1700s. In particular it became an important crop on the coast (sea islands) of the southern British colonies in North America.

Cotton is a very desirable fiber, but it required substantial hand work to remove the seeds from the fibers[115], comb the fibers straight and spin into yarn. The yarn was then woven into cloth. Overall, cotton had many desirable properties both physically

---

[115] Eli Whitney (1765-1825) patented the cotton gin in 1794, but it took several years to adapt it to work properly on hybrid "up-land" that was extensively grown in the southern States.

and economically compared to wool and silk. Thus, cotton cloth was highly prized in a very large market (rich and poor) in Europe.

The steps in processing cotton fiber (spinning and weaving) were obvious targets for automation in the 1700s and water-powered machinery in Britain brought spinning and weaving out of homes and into factories. There were several social effects: (i) Many people who had earned livings in spinning and weaving jobs were displaced with economic hardships; and an anti-industrial backlash followed (in many cases violent). (ii) Young women and boys were capable of many of the factory tasks and were induced to take jobs in these spinning and weaving factories under brutal conditions (long hours, dangerously unguarded moving equipment). (iii) The industry moved from London to Manchester (better sources of water power) and the port of Liverpool became the major import point.

The Trans-Atlantic Slave Trade

In the 1700s, the cotton technology was introduced to New England where water power was readily available to power factories. The New England colonies (Massachusetts/Main, Rhode Island, Connecticut, and New York) had many of the virtues and vices of Liverpool and Manchester. All were deeply involved in the trans-Atlantic slave trade initially for sugar plantations and later for cotton plantations. *Please note that these are two entirely different enterprises.*

In the 1600s and 1700s, the trans-Atlantic slave trade had focused on sugar grown almost exclusively on the islands of the West Indies. The nature of the sugar business required strong men to cut and process the sugar cane and the cane grew year-round. The islands (e.g., Hispaniola (French and Spanish), Cuba (Spanish), Jamaica (British), Antigua (British)) were disease-ridden (malaria, yellow fever, cholera). Thus, *millions of young African males* were bought in African, relocated to the islands, and worked to death *on sugar plantations*. These slaves never retired, they died within a about 5 years and created a steady demand for replacements. This model of slavery (no children and no retirees) was economically profitable…but obviously inhumane. In many cases New England families owned slaves for domestic chores. Tobacco plantations in the middle seaboard had acquired a few slaves from British and New England traders. But many indentured Europeans were available through the 1600s and these were pushed to the western frontier of the colonies as their contracts expired. Slavery per se (perpetually bondage) was gradually instituted and African slaves began to displace indentured servants in the late 1600s.

When cotton was found to grow well in the southern British colonies, it was *incorrectly* assumed that slavery would be a perpetual money-making process there as well. Thus, slavery was introduced to the southern colonies. But the crop was cotton; not sugar. Whereas sugar grew continually in the islands, cotton and tobacco, which could be raised in the southern colonies (Virginia, the Carolinas, Georgia) are seasonal crops. Moreover, the highly seasonal weather meant that labor has to be diversified. The labor in tobacco and cotton is less physically demanding that sugar and there is more skilled handwork. Thus, African women were commonly employed and very quickly there were African families. The absence of endemic tropical diseases meant that everyone was likely to live longer in the southern British colonies than in the West Indies.

New England and Britain had profited substantially in the international slave trade (i.e., bringing slaves to the West Indies and South America), invested their profits in water-powered cotton mills in New England. They brought slaves to the southern colonies assuming that profits would continue. But the slavery model that was active in the sugar industry did not hold in the cotton business.

The cotton gin (used to separate the seeds from the fiber) was introduced in the late 1700s. Thus, the relatively few slaves (about 500,000) brough to the southern colonies and states before 1807 (when the international slave-trade trade was outlawed) were in a completely different environment than those unlucky ones in the West Indies or in South America. Within 30 years there were African children in North America and the older slaves did not die of tropical disease and malnutrition… they retired. Moreover, they had to learn a variety of skills as the work changed with the seasons.

The population for Africans in the American south grew to about 4,000,000 in 60 years (1790-1850) …doubling at a rate of once every 25 years (1780-1860).

The first states to realize that there was a labor surplus were Virginia, North and South Carolina. These States experienced the fastest population growth, their tillable land was populated, and their unfertilized soils were worn out by the 1830s. In the 1830s, the Federal government displaced the Native Americans of the south to Oklahoma and auctioned off their lands to raise money. These new territories (northern Louisiana, Mississippi, Alabama, western Tennessee, Arkansas, eastern Texas) opened new ground for cotton. New production (especially in Mississippi and Alabama)[116] was not matched by new demand and prices dropped. By the 1860s, the economics of plantation cotton farming (which had never been nearly as profitable as the slave trade[117]) was clearly declining.

In the meantime, the potato famine in Ireland and northern Europe (1845-1860) had brought waves of immigrants via the ports of New York, Boston and Philadelphia into the upper mid-west. Coal-fired steam engines had been introduced and it looked as though the economies of the southern states was about to industrialize with no need for slave labor. But that did not

---

[116] Alabama statehood in 1819 and Mississippi statehood in 1817.

[117] Many large southern plantations were deeply in debt to British and New England cotton merchants who loaned money on assumption of future crops. By 1860, this market was in trouble because Manchester, England had stockpiled nearly a two-year supply of raw cotton in warehouses. The magnitude of this oversupply was not known or understood in the Southern States that assumed that they had substantial leverage over the British economy. They did not.

happen. [118] The American Civil War 1861-65 decimated the southern states and long-lasting social stresses were introduced causing racial problems that exist to this day. It is relevant that by 1870, *without slavery*, cotton production in the South exceeded levels of 1860. In many cases, child labor, replaced slave labor in the fields of the South.

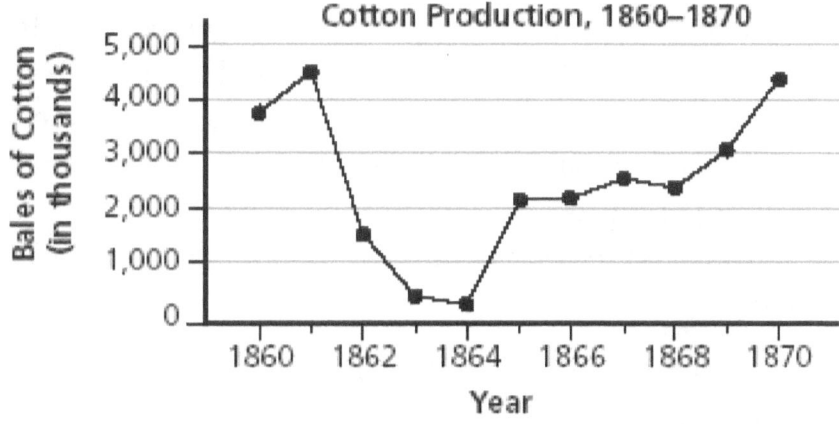

Source:

http://www.phschool.com/atschool/california/webcodes/itext_samples/in dependence/ebook/products/0-13-133479-4/gr8_ch12_pg441_1.gif

---

[118] New England had benefited from the proximity of hydropower (waterfalls) near its sea ports which facilitated the growth of the textile industry there. In the South, the fall line was not as dramatic and was often hundreds of miles inland from ocean ports. But steam power would allow industrialization in the South and coal could have been easily obtained from the Ohio Valley via the Ohio, Mississippi and Tennessee Rivers.

Nitrocellulose and Guncotton

The hydrogen bonding among cellulose strands can also be broken down by forming esters. This was discovered first with the use of nitric acid to form nitrocellulose. Notice that there are three R-OH units per carbohydrate unit. The more of these that are converted to esters $R-ONO_2$ the less hydrogen bonding there is. Thus, the cellulose becomes less ridged and more soluble. If the average number of nitrate esters exceeds 1.5, the product becomes easily flammable; and if the ratio goes above about 2.5 nitro groups per glucose unit the product can catch fire if heated above 150°C. And the highly nitrated product can be used as an explosive.

The history of this process begins in 1832 when Henri Braconnot added nitric acid to cellulose. Several other scientists followed during the 1830s, but the products are generally not stable. It was not until 1846 when Schönbein and Böttger essentially developed a process where the esters were formed by a brief exposure to a mixture of nitric and sulfuric acid and (importantly) the excess acid was washed away before the product was allowed to dry at low temperature (40°C).

$$H_2SO_4 + HNO_3 \leftrightarrows H_2O + HSO_4^- + NO_2^+$$

$$R-OH + NO_2^+ \rightarrow R-ONO_2 + H^+$$

Guncotton (trinitrocellulose) can be similarly formed by longer exposure to the nitrating solution at 0°C. This material is rather unstable. Guncotton has many advantages over black powder: it produces more gas, less heat and little smoke. The process was refined and patented by the firm of John Hall & Son (1846) who set up a plant in Kent (southeastern England) but the plant blew up killing a number of employees. These experiences were part of the background for the concern by Ascanio Sobrero as he advised against experiments with nitrate ester (e.g., nitroglycerine). The explosive uses were postponed until safer manufacturing technique were developed in the 1860s.

In 1869s, nitrocellulose was compounded with camphor and alcohol by John Wesley Hyatt, an Albany, N.Y., to produce a plastic that was used to manufacture billiard balls. It is said that these balls sometimes exploded when being struck.

Later nitrocellulose was used to make flexible film (celluloid). This film was used to carry photosensitive chemicals used in photography.

Pulp and Paper

There are three main components of wood: cellulose, lignin and sap. The balance of these components and details of lignin and sap vary from species to species of plant/tree.

Paper can be made by breaking up wood fiber in many different ways. In particular, simply grinding wood into small bits, extracting the sap and bleaching wood can be used to make paper. But to make really good paper or rayon, the cellulose needs to be freed of sap and lignin by chemical processing (called pulping).

Lignin seems to have evolved incidental to other metabolic activities of plants and single cell organisms. Plants synthesize a number of small molecules with 5 to 10 carbon atoms. One family of these molecules is based on cinnamic acid and coumaryl alcohol.

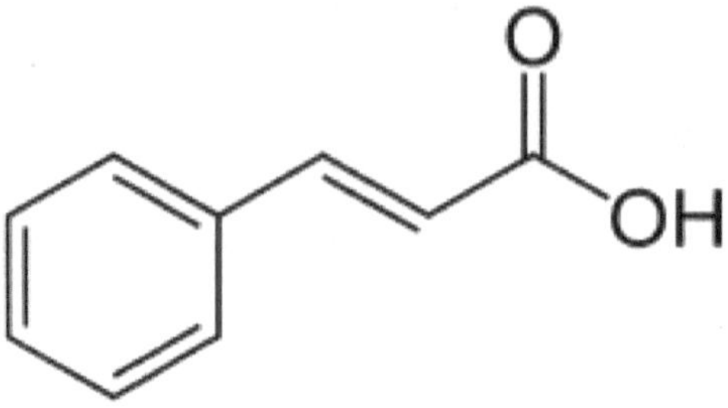

cinnamic acid

Author: NEUROtiker; Source Wikimedia Commons

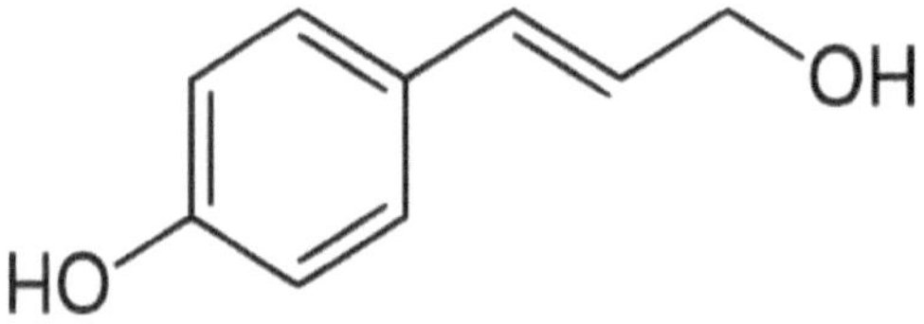

coumaryl alcohol

Author: Yikrazuul; Source Wikimedia Commons.

Coumaryl alcohol is an intermediate in the biosynthesis of many other compounds including:

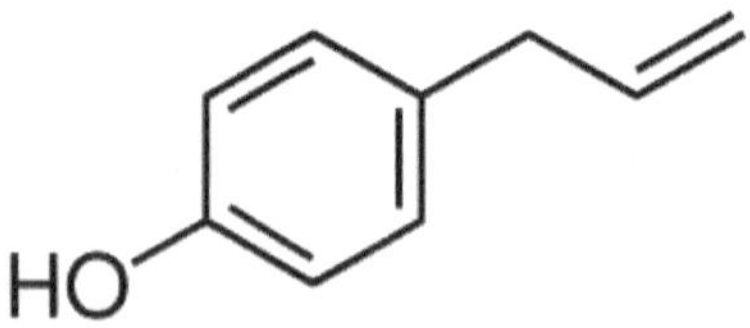

Chavicol

Author: Calvero; Source Wikimedia Commons

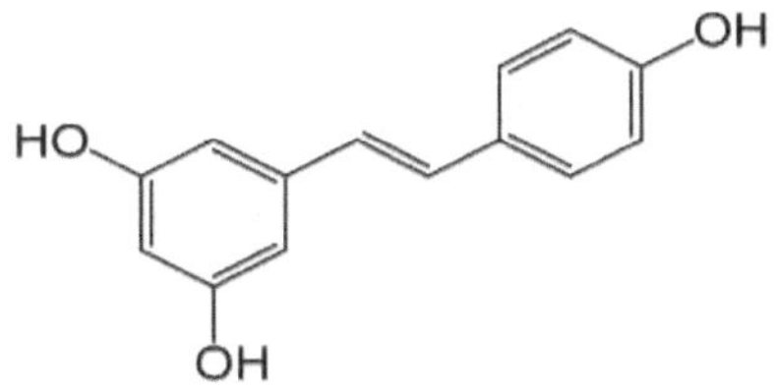

stilbenoids

Author: Fvasconcellos; Source Wikimedia Commons

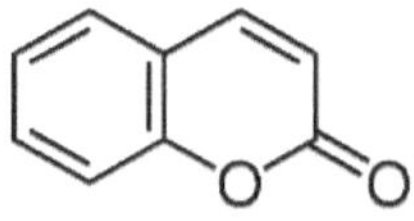

coumarin

Author: Calvero; Source Wikimedia Commons

Other members of the coumaryl family have one or two methoxy groups adjacent to the phenolic hydroxide.

One of the main functions of coumaryl alcohols in cells is as a reducing agent to suppress free O_2 and/or hydrogen peroxide that leak from chloroplast and mitochondria in the cytoplasm to prevent damage to the DNA and RNA. They are hydrogen donors for oxidase and peroxidase enzymes, which leaves them as moderately stable phenolic free radicals. These radicals diffuse to the outside of cells and undergo random coupling reactions. The polymer that is formed is what we call lignin.

Lignin has no fixed structure and contains a large variety of sub-structures with C-C and C-O bonds between the monomers that make it up. In some cases, it may include covalent bonds to proteins and carbohydrates. When plants die and fall to the earth, biological and abiotic processes quickly consume the carbohydrate and protein of the plant as well as many of the small molecules. The lignin is partially oxidized (aliphatic alcohols are converted to carboxylic acids and ketones and phenols to quinones. The result is what we call "soil organic matter," i.e., humic acids that are partially soluble in base.

The most tractable structures in lignin that occurs in significant amounts are variations on this:

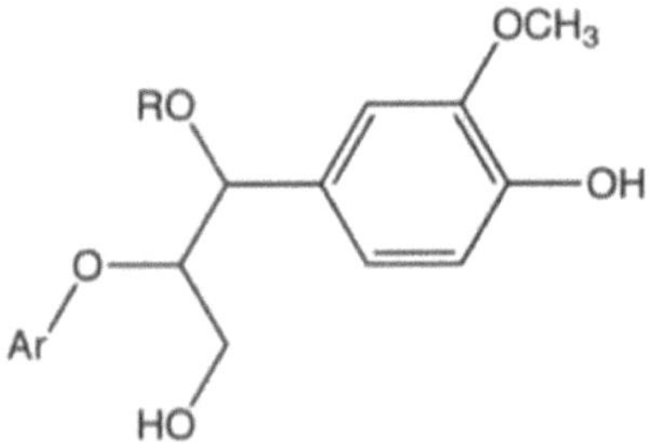

Where the R and Ar symbols represent aliphatic and aromatic linkages within the extended polymer. Separating the lignin from the cellulose can be accomplished by breaking these linkages to cut the lignin into smaller pieces. This is done in the Kraft pulping process:

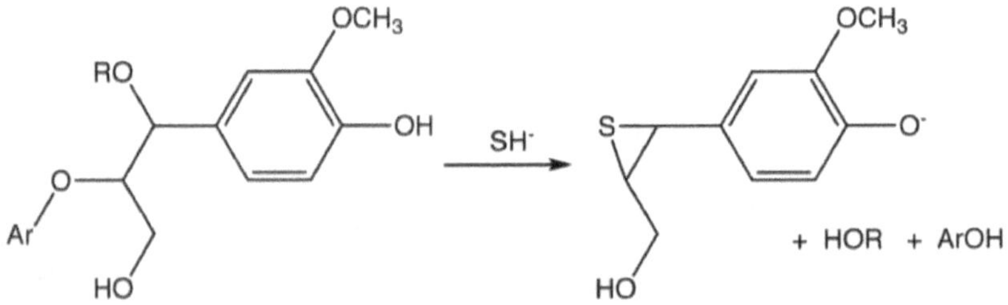

Author: Smokefoot; Source: Wikimedia Commons

The Kraft process began in the early 1800s and is noted for the fact that it eliminated the lignin and hemicellulose (amorphous carbohydrate polymers) without degrading the cellulose. As a result, the cellulose polymers are long and produce strong paper. Processes that use acid (e.g., sulfurous acid) break the lignin in a similar fashion but degrade the cellulose (as well as the lignin) producing inferior paper.

Most of the developments over the years involve how to optimize the overall process to make it fast and efficient. Very briefly, wood is chipped to a fine size to expedite penetration by the chemical reagents. The chips are heated and steamed to remove pockets of gas and make them soft and flexible. Then the essential chemicals called "white liquor" consisting of sodium hydrogen sulfide (sodium hydroxide and hydrogen sulfide) are pressured into the chips. The mixture is "cooked" at about 175°C for about 1.5 hours. It is important that the temperature not be higher to prevent degradation of cellulose (the actual cooking time is dependent on the size of the chips). The sulfide ions are excellent nucleophiles and break the lignin into soluble fragments as shown above.

The entire "cook" is then washed and filtered to extract as much of the lignin as possible leaving a tan colored mat of cellulose fibers. The liquid is now black and is called "black liquor." It is concentrated by evaporation of water and passed into a burner where the carbon compounds are combusted leaving the minerals, which are recycled (green chemistry).

The pulp (cellulose) is sent through several washing and bleaching stages. The pulp can be used in manufacture of paper, rayon or other products.

Rayon

It is not surprising that a variety of research activities in the late 1800s centered around cotton, which was one of the fibers of most interest to dye chemists. Paul Schützenberger (1829–1897) is credited with discovering that acetic anhydride reacts with cellulose to form esters analogous to the nitrate esters (1865). But it was not until 1910 that Camille Dreyfus (1878–1956) successfully made an acetate film that compared to the nitrate film being used in photography. The principle was the same: Break down the hydrogen bonding and the cellulose becomes syrup. The acetate film began replacing the highly flammable nitrate film and a lacquer produced by diluting the acetate with solvent found a market in the fledgling airplane industry as a coating for cloth used to cover the cloth surfaces of wooden airplanes. The development of cellulose acetate as a fiber in the 1900s was hampered by the thermal instability (both melting and decomposition).

Matthias Schweizer (1818–1860) discovered that cellulose would dissolve in a solution of copper sulfate/ammonium hydroxide. It was another variation on the same theme. In this case, the copper amine complex would bind two of the hydroxyl groups (on carbons 2 and 3) as a chelating ligand:

$$\text{-CHOH-CHOH-} + Cu(NH_3)_4^{++} \rightarrow R(O)_2Cu(NH_3)_2 + 2\ NH_4^+$$

This complex could be extruded into an acid solution and the insoluble cellulose was regenerated as a filament.

The method has had the most commercial success was first patented by Cross, Bevan and Beadle in 1894. The method involves using a strong base to catalyze the reaction of cellulose with carbon disulfide.[119] The product is a xanthate:

$$ROH + {}^-OH + CS_2 \rightarrow R\text{-}OCSS^- + H_2O$$

analogous to a carbonate. This solution is extruded through spinnerets into an acid solution, which reverses the reactions and returns the cellulose.

Rayon fibers were advertised as "artificial silk," but did not hold up well to washing and drying. Of course, the nitrate versions were downright dangerous. It is not clear but partially nitrated cellulose may have been involved in a series of heartbreaking clothing fires of "torch sweaters" between 1947 and 1953. An interest in fire protection was cited in the relocation of the National Bureau of Standards (now NIST) to Gaithersburg, MD. Congress passed the *Flammable Fabrics Act* in 1953.

It was common to find viscose rayon plants virtually collocated with pulp and paper plants because the sulfite pulping process is able to produce cellulose from wood that is compatible with the manufacture of rayon.[120]

---

[119] Carbon disulfide is very volatile and has a very low autoignition temperature. It will catch fire on a hotplate and burns with a cool blue flame. It also has a nauseating odor.

[120] I recall vividly the period around 1960 when my family's summer vacation would take us through Canton and Inka neighboring towns in

Tall Oil

Pulping plants process enormous amounts of wood and naturally there is some sap associated with the wood (~40 kg / ton pulp). This oil collects in large enough quantities to be separated from the black liquor and has been used as a biofuel. However, there is interest in recovering some of the chemical components as chemical intermediates. These include fatty acids and abietic acid ($C_{20}H_{30}O_2$):

Author: Edgar181; Source: Wikimedia Commons

Lower molecular weight compounds found in sap, such as pinenes (terpenes) appear to be destroyed or converted into water soluble compounds in the pulping process.

---

the mountains of North Carolina. The stench of carbon disulfide permeated one town and hydrogen sulfide filled the other.

Turpentine

The sap of pine trees can be collected and processed to yield lower molecular weight hydrocarbons composed principle of alpha and beta pinene:

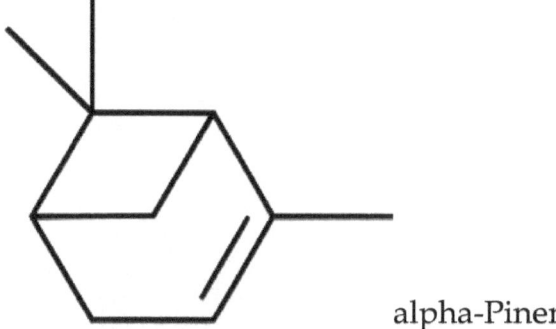

alpha-Pinene ($C_{10}H_{16}$)

Author: J.delanoy; Source: Wikimedia Commons

Some closely related compounds such as limonene provide distinctive aroma to citrus fruits.

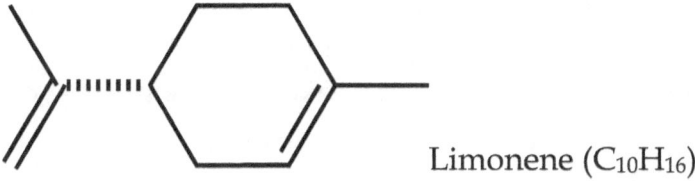

Limonene ($C_{10}H_{16}$)

Author: Karlhahn/ Benjah-bmm27; Source: Wikimedia Commons

Natural and Synthetic Rubber

You may have notices that many of the chemical structures shown in the last few pages could be produced from C_5H_8 fragments.

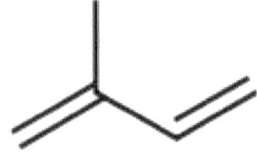

Isoprene (C_5H_8)

Drawing by Fvasconcellos, source Wikimedia Commons

Natural rubber was discovered by the Olmec culture in central America some time before Christopher Columbus's second voyage, because he returned to Europe in 1496 from the West Indies with samples of the material. Ut is likely that some sort of vulcanizing process was invented in Central America; but its secret was not carried to Europe. It was first described by Europeans in 1525, but it was not subjected to serious study until 1735 by Charles de la Condamine. Priestley discovered that it would erase writing and that is where the English name came from. In 1815, rubberized canvas was produced by dissolving rubber in turpentine (later benzole), soaking canvas with the liquid, and evaporating the solvent. This material was employed in waterproof boots and clothing (e.g., the MacIntosh).

To this point, rubber lacked strength and became brittle at low temperatures. In 1840, Charles Goodyear (1800–1860) discovered that heat and sulfur made rubber more durable, but did not immediately patent it. However, in 1842, Thomas Hancock (1786–1865) of the MacIntosh firm acquired a sample of Goodyear's rubber; discovered how Goodyear had made it and patented "vulcanization." Vulcanized rubber immediately found important markets such as toys and the pneumatic tire (1845), which was widely utilized in the early bicycles (1869).

Throughout this period, the natural source of rubber was very difficult to grow and harvest. Apparently in the 1700s, the British transplanted a few trees from Central America to South America[121], India, Africa and Southeast Asia. But it took long periods to grow the trees and harvesting the latex was literally done drop by drop. Obviously, interest in the composition of rubber and methods of its synthesis grew in Europe.

In 1860, Charles Greville Williams (1829–1910)) destructively distilled natural rubber and obtained a light oil (b.p. 37-38°C), which has the same composition as natural rubber (88.23%

---

[121] Some sources indicate that rubber trees were native to Brazil. The Olmec center (El Manati) is in southern Mexico, which is roughly 2,500 difficult miles from Manaus, Brazil. Thus, it seems unlikely that rubber trees were growing in Brazil before the Europeans brought them there. I see no mention of rubber in Manaus, before the mid-1800s.

carbon and 11.77% hydrogen) and which he called "isoprene" (C_5H_8).[122]   However, the structure of isoprene was not understood until 1882 through the work of William Tilden (1842–1926).

In the late 1800s the demand for natural rubber expanded and there was a rubber boom in Brazil (1879 to 1912). This boom had a generally negative impact on Brazilian natives who were in some cases reduced to slavery; much the same way that natives of the Congo basin in Africa were experiencing a similar involvement with Europeans.[123] Indeed, slavery did not end in Brazil until 1888. In the meantime, Europeans became very wealthy in the Amazon. After slavery ended in Brazil, the rubber trade shifted to the East Indies. British entrepreneur Henry Wickham (1846-1928) shipped upwards of 100,000 seeds from rubber trees in Brazil to the London botanical gardens as "scientific specimens," which would normally imply that they were dead. But about 3% of these seeds germinated and were shipped to British estates in Ceylon and Malaysia. From there, they spread into the Dutch East Indies. Ironically and probably foolishly, Henry Ford attempted to reestablish rubber in Brazil (in as much different social climate) in the period 1928-1934. But

---

[122] Isoprene does not automatically form rubber. It was not successfully polymerized until 1955 by Samuel Horne of the B. F. Goodrich company.

[123] Portuguese in Brazil (Amazon basin) and Dutch in the Congo basin.

the project was abandoned because it could not compete economically with rubber from the East Indies.

Thus, the East Indies became virtually the world's sole source of natural rubber after 1934. When Germany overran western Europe in 1941, the Dutch Royal family evacuated to Britain such that the British were essentially the protectorate of the Dutch East Indies. That ended on December 7, 1941 when the Japanese neutralized the American forces in Hawaii, Wake Island and the Philippines opening the door to acquire the European (British, Dutch, French) assets in the Far East. These assets included petroleum, latex/rubber, quinine and tin. Rubber and quinine were uniquely produced there and these loses initiated major research programs in the United States.

In Germany, I. G. Farben (Bayer) had developed a polymer of butadiene and styrene called Buna-S by 1929. This was commercialized by 1938 and provided the synthetic rubber for Germany throughout the war. Butadiene is manufactured from the light butane/butene hydrocarbons from petroleum and synthetic oil production by dehydrogenation using the Houdry process:

$$C_4H_{10} \rightarrow H_2C=CH-CH=CH_2 + 2\ H_2$$

Styrene is obtained from ethylbenzene (which can be isolated from *benzol* or synthesized from ethylene):

$$\text{(ethylbenzene) } C_6H_5-CH_2-CH_3 \rightarrow \rightarrow C_6H_5-CH=CH_2 \text{ (styrene)}$$

Butadiene and styrene are polymerized using a peroxide catalyst.

Although rubber tires come immediately to mind as an essential war commodity, the air forces of the world were also dependent on a variety of rubber parts and especially on self-sealing fuel tanks. Apparently, there were three approached to self-sealing fuel tanks taken during the war. The Americans put the rubber materials inside the metal fuel tank and this ultimately facilitated use of the wing sections themselves as fuel cells. This option was facilitated by American uses of largely aliphatic high-octane fuel, which had low solvent power (*e.g.*, does not attack rubber). The British and especially the Germans were forced to use high aromatic fuels (*benzol*) to achieve the octane-rating they needed and this appears to have led them to place the rubber self-sealing materials outside the steel fuel tank. The Japanese, to minimize weight and maximize range, simply did not use self-sealing fuel tanks.

7.3 Formaldehyde Resins

Formaldehyde ($H_2C=O$) was first accidentally produced and identified by its characteristic odor in 1859. But it was not systematically produced until 1868 by A.W. Hofmann by oxidation of methanol (H_3COH) over a glowing platinum wire. Over the years, the method has been improved to avoid side reaction (e.g., decomposition to CO and H_2, excessive oxidation to formic acid (HCO_2H), and ought-right explosions to CO_2 and H_2O).[124] The first commercial manufacture was in 1889 using a copper gauze catalyst. A silver oxide catalyst was introduced in 1910. The production was limited in the early 1900s because the methanol was obtained from distillation of wood ("wood alcohol"). The development of a high-pressure methanol synthesis (from synthesis gas) in 1925, allowed great expansion in the preparation of formaldehyde.

 Formaldehyde a very reactive gas and is generally obtained as an aqueous solution with methanol. It readily forms addition products in these solutions:

$H_2C=O + H_2O \leftrightarrows HOCH_2OH \leftrightarrows$ polymers (including trioxane)

$H_2C=O + H_3COH \leftrightarrows H_3COCH_2OH$ (hemiacetal)

$\leftrightarrows H_3COCH_2OCH_3$ (acetal)

The safety of formaldehyde products has come under scrutiny as formaldehyde is a volatile compound. The same chemical reactivity that is of interest in commercial applications (see below), are potentially very adverse *in vivo*. Formaldehyde is expected to react with proteins, carbohydrates, DNA and RNA. The result varies from irritation of the membranes to shut down of biochemical systems to cancer.

---

[124]

https://www.ukessays.com/essays/environment/formaldehyde.php

Urea-Formaldehyde Resins

It was recognized early in its history that formaldehyde will react with ammonia and amines. Indeed, hexamethylenetetramine was one of the first compounds made with formaldehyde.

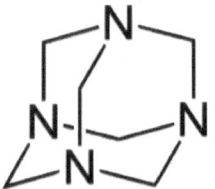 Author: Yikrazuul; Source: Wikimedia Commons

The reaction with urea (H_2NCONH_2) provides a "resin." In general, the term "resin" denotes a partially polymerized material with active "ends" that can be completed as an inert polymer by "curing" (typically by heat or catalyst):

Author: Smokefoot; Source: Wikimedia Commons.

Urea-formaldehyde resin is very compatible with wood products (cellulose polymers) and with heat and pressure these resins are used to glue plywood sheets together. Plywood is made by literally peeling "sap wood" from pine logs with a sharp knife leaving behind the "heart wood."[125] The thin sheets of sap wood are laminated with urea-formaldehyde resin adhesive with the natural grain at different angles to provide strength and stability.

There have been several episodes of "trailers" (mobile homes) provided for temporary emergency housing in the US producing harmful levels of formaldehyde in the interior air. These poorly ventilated shelters may be experiencing reversal of the polymerization process in a hot/humid environment where fungus may play a role in rotting the plywood or wooden structures glued together with formaldehyde-based adhesives. It is not clear whether or not the materials were properly "cured" in the first place.

Love Canal

In 1889, Leo H. Baekeland (1863–1944) immigrated to the United States from Belgium and continued his interest in photographic chemicals. By 1893 he had established a company with partners and developed a process for making photographic papers. The company was sold to George Eastman (Eastman Kodak Co. of Rochester, NY) in 1899 making Baekeland a wealthy man, but the

---

[125] The heart wood "peeler cores" are typically pressure treated with preservative fungicide and sold as to fence post. But, because the preservative does not penetrate the heart wood well, these posts typically rot at the ground line in about 5 years.

deal included terms that prevented him from producing photographic products of 20 years. About 1890, Baekeland travelled to Germany and studied electrochemistry.

In 1892, William T. Love obtained funding to build a seven-mile long canal to divert part of the water from the Niagara River from above the falls to below the falls. This would have been an important source of hydroelectric power and Love envisioned an important city growing on the site. Unfortunately, after digging only about 4,600 feet, the country went into a recession and the project was terminate. The site was largely abandoned and the canal was used for recreation for many years.

Returning to Niagara Falls, New York, Baekeland consulted for Elon H. Hooker as he established the Hooker Electrochemical Company (1903), which established a chlor-alkali plant to make chlorine and sodium hydroxide solution which are the ingredients of sodium hypochlorite bleach (1906) and "chlorinated lime" (calcium hypochlorite, solid $Ca(OCl)$).

$$2\ NaCl + 2\ H_2O \rightarrow Cl_2 + 2\ NaOH + H_2$$

Phenol was originally obtained commercially by extraction form coal tar, but as the market grew, several companies developed methods for making phenol from benzene. These processes and electrophilic substitution on benzene with either sulfuric acid (SO_3) or chlorine, followed by high temperature and pressure nucleophilic displacement of the substituent with hydroxide.

One of the first ventures of the Hooker Company into organic chemistry was to manufacture chlorobenzene for conversion to phenol used during World War I (1914-1918). Phenol was used to make picric acid (trinitrophenol), which is an explosive that was used in artillery shells.[126] Other organic chemicals followed including chlorinated solvents.

Meanwhile, in the 1920s, the municipal government of Niagara Falls began dumping waste in Love Canal. During WWII (1940s) many plants were build in the area and the federal government appropriated the site as a dumping ground. Hooker had acquired the site; and in 1942, they drained and lined part of the canal with clay to dispose of chemicals they knew would be undesirable to release in an uncontrolled fashion. The company used the site for waste disposal until 1953. When this use was ended, the site was covered with soil and soon looked like an ordinary empty field in an area that was experiencing an economic boom. The Niagara Falls City School District wanted land to build new schools. When Hooker advised against that use and resisted sale the local government forced the sale of the land by invoking eminent domain. Hooker made a clear disclosure of the nature of the site and donated it to the government for $1.00.

This is not the last we will hear of Love Canal.

---

[126] Dow and Monsanto also made phenol for picric acid. Meanwhile in Germany, Fritz Haber turned his attention to production of chlorine as a chemical weapon that was used in the first modern chemical weapons attack at Ypres April 22, 1915.

Bakelite: Phenol-formaldehyde Plastics

Meanwhile, Leo H. Baekeland had become the father of the plastics industry. The -OH group on an aromatic ring makes the ring susceptible to electrophilic attack especially at the *para* and *ortho* positions:

para-   *ortho-*

Source: http://www.chem.ucalgary.ca/courses/351/Carey5th/Ch24/ch24-3-1.html

As noted above formaldehyde is an excellent electrophile and various people had reacted phenol and formaldehyde obtaining gooey compositions that were discarded as useless. Baekeland patiently and carefully studied this reaction over a period of years carefully noting the effect of composition, temperature, concentration etc. on the resulting product. His first products were a series of soluble resins, which he called Novolaks (or novolacs). These compounds are prepared with less than one mole of formaldehyde per mole of phenol. They were marketed as a form of shellac that was cured with various agents (acids or amines).

He eventually found a set of conditions that produced a solid resin (small particles) that could be placed into a mold and cured by heating and pressure. He filed a patent in 1907 and it was granted in 1909. Of course, this greatly expanded the market for phenol and formaldehyde. *Bakelite*, as the plastic was called, is a random three-dimensional polymer of phenol molecules joined primarily at ortho and para positions by methylene ($-CH_2-$) units and ether units ($-OCH_2-$). Thus, the material does not melt without decomposition and is very durable and rigid. It is also a good insulator for heat and electricity. It found wide application in the expanding electrical, automotive and aeronautical industries that were expanding during and immediately after the First World War.

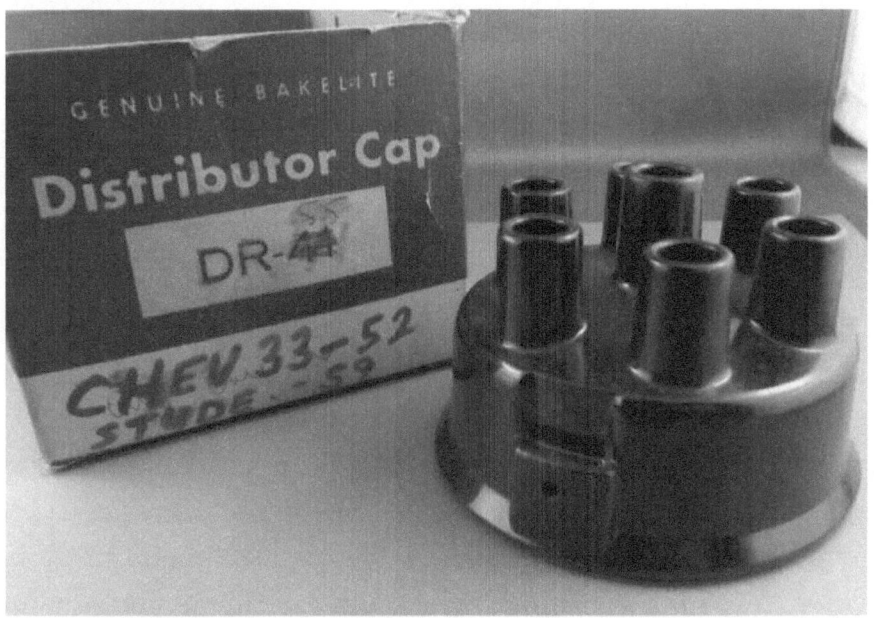

The distributor cap (used on gasoline engnes) is perhaps one of the most iconic examples of Bakalite technology.

7.4 Epichlorohydrin and Epoxy Resins

Acetone

Acetone is a carbonyl
compound similar to
formaldehyde, but the
carbonyl is much less reactive.

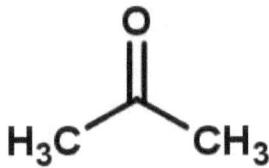

Acetone was identified in the 1800s and had few significant uses
until it was discovered to be useful in the manufacture of cordite
(a military explosive). Cordite[127] was a low-velocity explosive
that replaced black powder in the late 1800s for military
applications because it is smokeless. Acetone was used to soften
cordite so that it could be extruded into "cords" (thick filaments)
which could be broken into pieces of various sizes for filling
carriages (1889).

Chaim Azriel Weizmann (1874–1952) was a biologist who moved
to Britain in 1904 and developed fermentation as a method of
producing industrial chemicals. Of course, the idea of making
ethanol from starch had been practiced for a very long time, but
Weizmann's idea was to select a specific micro-organism and
provide it a specific starting material to obtain a useful product.
He discovered that he could make acetone, butanol and ethanol
(easily separated by distillation) simultaneously from starch with

---

[127] Weight percent: 58% nitroglycerine, 37% nitrocellulose, 5%
petroleum jelly.

Clostridium acetobutylicum. The need for acetone was recognized early in the war and by 1916 Weizmann's process was producing acetone in a number of breweries. A shortage of starch (from corn imported from the US) developed as German U-boats attempted to blockade Britain. A call went out for school children to search the woods for horse chestnuts and acorns (October 1917) as a source of starch for the Royal Naval Cordite Factory.

The Cumene Process

The modern process for making both phenol and acetone was developed in the 1940s and has been refined over the years. It involves an electrophilic alkylation of benzene with propene. By careful control of the catalyst and reaction conditions, monoalkylation can be achieved to produce isopropylbenzene (a.k.a., cumene). It turns out that the hydrogen at the central position of the propyl group is very easily removed (because the phenyl ring and two methyl groups stabilize the corresponding free radical). Thus, oxygen (O_2 which has unpaired electrons) is able to extract the hydrogen atom and form a hydroperoxide under ambient conditions. The hydroperoxide rearranges by shift of the phenyl group and then decomposes to phenol and acetone.

Similar reactions with butene produces butanone. If the diisopropylbenzene is prepared, it can transform to the diphenol (hydroquinone), which can be oxidized to 1,4-benzoquione.

Synthetic Glycerin

The invention and wide-spread use of dynamite (nitroglycerin) in the late 1800s, had already made glycerin a high priority industrial chemical. Saponification of fatty acid esters was providing most of this market. But separating glycerin (bp 290°C) from aqueous hydroxide solutions was expensive.

When a source of propene from petroleum refining became available, allyl chloride could be made by free radical chlorination (gas phase) and chlorination of ally chloride in caustic solution produced a mixture of dichloropropanols.

$$Cl_2 + H_2C=CHCH_3 \rightarrow H_2C=CHCH_2\text{-}Cl + HCl$$

$$H_2C=CHCH_2\text{-}Cl + ClOH \rightarrow Cl\text{-}H_2CC(OH)HCH_2\text{-}Cl$$

Refluxing this mixture eventually displaces the chlorides to produce glycerin. But it was discovered that in this process, epichlorohydrin (which can be easily distilled from the reaction mixture, b.p. 118°C) is formed

Author: Benjah-bmm27; Source: Wikimedia Commons

Epichlorohydrin can then be converted to glycerin.[128] When epichlorohydrin became a major commodity, research began on what it could be used for.

Bisphenol A

If you attempt the same reaction that produces Bakelite resins substituting acetone for formaldehyde, the principle product is "bisphenol A."

Author: Darkness3560; Source Wikimedia Commons

In addition to the para, para substituted product, there is also a para-ortho compound and a cyclic condensation product.

Part of the interest in bisphenol A stemmed from its structural similarity to estrogen.

---

[128] It should be noted that with the recent extensive interest in biodiesel fuel from the hydrolysis of fats over the last few decades, the economics have reversed and now epichlorohydrin is prepared from glycerol rather than from propene.

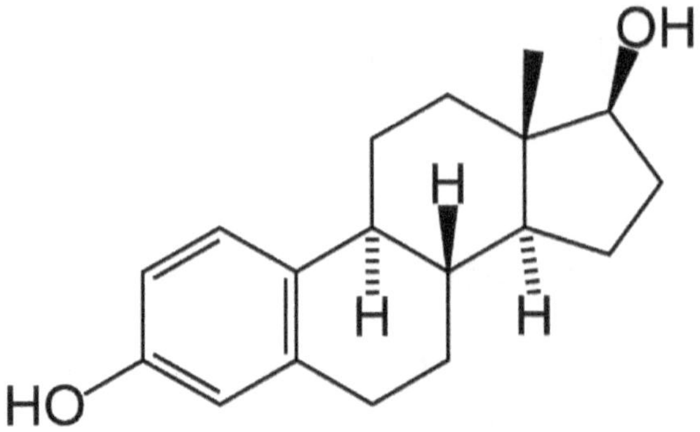

Estrogen E2

Source: https://menopausehealthmatters.com/progesterone-deficiency/estrogen-dominance/

Bisphenol A has much less estrogenic activity that estrogen (1/37,000) as tested by Edward C. Dodds (1899–1973) in 1936.[129] But diethylstilbestrol is an estrogenic compound that was used therapeutically until it was found that it caused cancer.[130]

---

[129] Dodds and Lawson. *Nature* 137:96 (1936).

[130] The carcinogenic effects may be related to fragmentation of mitochondria, which release high concentrations of reactive oxygen species into the cell and may cause chromothripsis.

Smoly JM, Byington KH, Tan WC, Green DE.
On the fragmentation of mitochondria by diethylstilbesterol. II. On the relation of the released proteins to the mitochondrial membranes. *Arch Biochem Biophys.* 1968. 128(3):774-89.

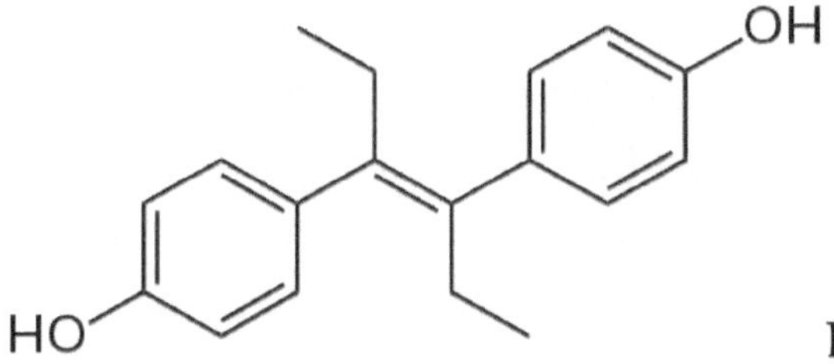

Diethylstilbestrol

Author: Kopiersperre; Source: Wikimedia Commons

More recently, the subtle role of estrogenic compounds ("hormone disrupters") has been considered an environmental issue.

Epoxy Resins

The first epoxy resins were prepared by reacting epichlorohydrin with alcohols and the Novolacs were ideal because the phenolic hydrogen is slightly acid. The reaction of a phenol with epichlorohydrin first opens the epoxy ring to form a 3-chloro-2-hydroxypropyl ether, which forms a new epoxy unit by displacement of the chloride by the adjacent hydroxyl group. The result is a glycidyl ether.

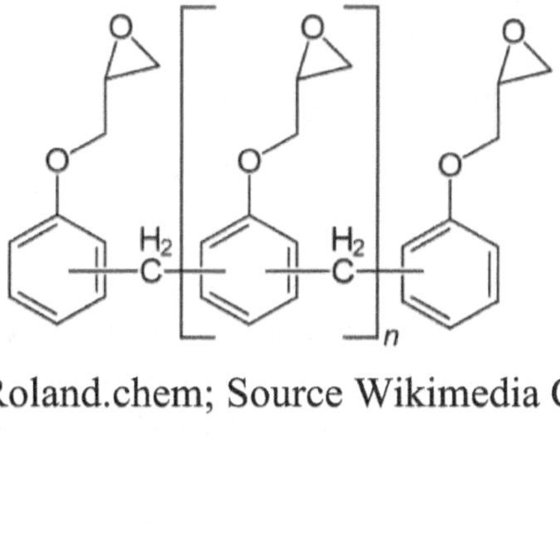

Author: Roland.chem; Source Wikimedia Commons

Page 335 of 492

In Germany, Paul Schlack made these glycidyl esters and cured them with amines (1934) to produce thermosetting polymers. Modern (high performance) epoxy resins began with the reaction of bisphenol A and epichlorohydrin in 1936[131] by Pierre Castan in Switzerland.

Author: Minihaa; Source: Wikimedia Commons

Polycarbonates

The bisphenol A backbone provides a stable structure and it is also used to make linear polycarbonates by reaction with phosgene ($Cl_2C=O$). The idea had been around since the late 1800s, but commercial products were not invented until 1953; and importantly crystal-clear material suitable for optical lenses was not available until 1970. In 1983, the first high index polycarbonate lenses for glasses were marketed (they were much

---

[131] Gannon J.A. (1986) History and Development of Epoxy Resins. In: Seymour R.B., Kirshenbaum G.S. (eds) High Performance Polymers: Their Origin and Development. Springer, Dordrecht

thinner and lighter than glass and more shatter resistant).
Polycarbonate (refractive index 1.58) is now (1987) outclassed by
polyurethane plastic lenses (refractive index 1.66).

7.5 Polyesters and polyamides

Proteins

Protein is a natural polyamide made up from amino acids. There
are 21 amino acids that have evolved with functional utility in
eukaryotes. They all have the general formula $H_2N\text{-}CHR\text{-}CO_2H$,
where the R group varies. In general, the -R groups can be
classified as ionic (positively charged or negatively charged),
polar/hydrogen bonding, or hydrophobic. As a result, the
polyamides formed from the amino acids (i.e., peptides) tend to
fold into preferred configurations (active proteins) in aqueous
solution.[132] Proteins can be confined to certain conformations by

---

[132] There have been attempts to predict the shape of proteins based on
the sequence of amino acids. The success of these systems seem to be
limited because (i) the peptide chain is formed in sequence and there
may be folding that occurs early in the process (involving only a few of
first amino acids), which may be unstable, but cannot be undone; (ii)

forming covalent cross-links with -S-S- (disulfide) linkages between cysteine units. These linkages are sensitive to the oxidation and reduction conditions in cells. The acid and base units are also sensitive to local pH conditions.

Some proteins are essentially structural (in particular, we are interested in wool, silk, and hair in this book) but many proteins are intended as functional biochemical tools. At this level, the proteins are capable for a myriad of tasks that are not feasible with conventional chemical reactions. In some cases, the proteins contain "active sites" where substrate molecules fit into and are quickly and selectively transformed. In other cases, the protein "active site" is a receptor for a messenger molecule, which causes a change in conformation of the protein causing (or blocking) a series of chemical reactions. Perhaps, even more interesting, are the proteins that act as motorized tools. In these cases, the "fuel" is provided by ATP that transfers a phosphate group into a hydroxyl group;

$$R\text{-OH} + ATP \rightarrow ADP + R\text{-OPO}_3{}^{2-}$$

These reactions are selectively catalyzed by special enzymes called "kinases." Notice that when the phosphate was added, the polar hydroxyl group was converted into an even more

---

eukaryote peptides include segments that are removed as the protein is formed, (iii) there may be template molecules that chaperone the peptide into its final shape, (iv) glycans (sugar molecules) frequently become attached to the peptides and appear to be important to guiding the protein to is needed position in the cell.

hydrophilic ionic phosphate group. This change can change the configuration of the protein. Removal of the phosphate,

$$R\text{-}OPO_3{}^{2-} \rightarrow R\text{-}OH + PO_4{}^{3-}$$

causes the protein to return to its original conformation. Thus, we have a reciprocating tool (e.g., opening and closing its jaws).

I have been fascinated by the topoisomerase enzyme. Let me explain what is does:

> Have you ever had a ball of twine, electrical extension cords, or section of garden hose that was twisted and knotted? As you struggle to find the ends and figure out how to pass the long strings through loops so as to untangle the mess, you usually think to yourself, "If I could only cut this string and put it back together as it was, I would be able to do this job easily." Well imagine the problem of DNA in a eukaryote cell (one molecule wide and hundreds of millions of molecules long). Tangled in thousands of ways. Topoisomerase, can grab a strand of DNA, pull it apart, pass another strand through it and rejoin the strand it broke perfectly as good as new.

I first saw that process in a diagram in the 1990s and immediately regretted not going into biochemistry.[133]

---

[133] When I took chemistry in college (1966-74), biochemistry was very vague about the shapes function of proteins. The terms were "peptide" and "protein" were as far as they went with occasional references to molecular weights. The idea of "molecular biology" was not even considered unto 1975 at the Asilomar Conference organized

Synthetic Polyesters and Polyamides

Wallace Hume Carothers (1896-1937) was hired by Dupont in 1928 from Harvard University. In my opinion (based on review of the situation); this was a bad idea for everyone. Carothers really had very little experience (he received his PhD in 1926 and had just moved to Harvard) and Carothers had serious psychological problems. He committed suicide in 1937 and Dupont seems to have attributed most of the work of others (colleagues and subordinates) to him in sympathy. Thus, I will not spend much time on Carothers.

Julian W. Hill (1904-1996) graduated from MIT in 1928 and was also hired by Dupont in 1928. He appears to have been the driving force behind polyester research at Dupont. Unfortunately, he was assigned to work under Carothers. His first efforts produced a polyester polymer but it was unstable in water. Thus, the team switched to polyamides, which were expected to be more resistant to hydrolysis and not soluble in water. The first nylon was made by Donald D. Coffman. It was decided to work with hexamethylenediamine and adipic acid, which could both be synthesized from benzene. When mixed in equal-molar amounts, they form an ammonium salt which can be

---

by Paul Berg (1926-). I attended the 1976 meeting in Washington, DC on behalf of the USEPA because DNA is a chemical substance as defined by the Toxic Substance Control Act. I strongly advocated that EPA not contest NIH leadership in this area and NIH soon issues its rules. Recombinant DNA research guidelines. *Federal Register*. 1976;41(131):27902–27943

isolated and pyrolyzed with the expulsion of water. The product could be melted and extruded as fibers (1934).

Nylon 66 Author: Dominik-jan; Source Wikimedia Commons

The compound was initially a commercial success as coarse fibers (tooth brush bristles) but soon was extruded in fine fibers that could be woven like silk. Indeed, the nylon stocking were introduced in 1938 and resulted in a series of "nylon riots."

https://www.slideshare.net/guimera/the-nylon-stocking-is-75-years-old

Caprolactam

The success of nylon has resulted in interest for more economical approaches to industrial synthesis. Today, most nylon ('nylon 6") is the polyamide produced from rearrangement of caprolactam at 260°C, which is conveniently prepared from cyclohexanone by the Beckman rearrangement:

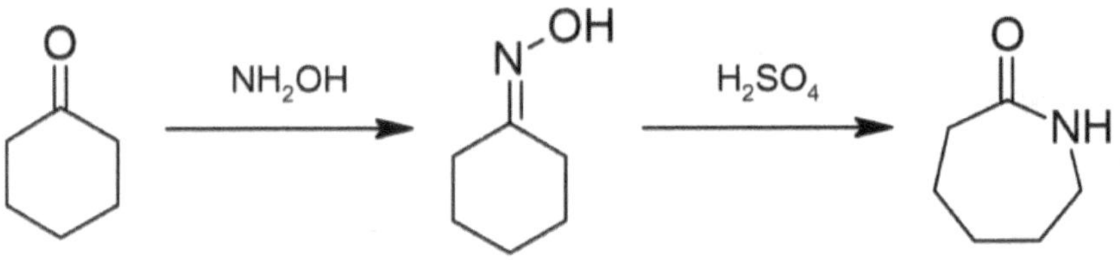

1   **2**   **3**

Author: Stone at English Wikipedia; Source Wikimedia Commons

Polyesters

The problem with polyesters produced in the 1930s was not stability of the ester linkage, but rather solubility in water. After the success of nylon, most American companies focused on polyamides. It was not until 1941 that polyethylene terephthalate was synthesized in Britain by John Whinfield (1906–1966) and James Dickson (Calico Printers Association). In the US, the product is known under the trade name Mylar™. The plastic is normally prepared from dimethyl ester of terephthalic acid and ethylene glycol with the methanol being distilled from the reaction mixture (acid catalyzed ester exchange).

Because the diacid is relatively expensive to manufacture and because the polymerization can be reversed under relatively high temperature, recycling of PET plastic is technically and economically feasible. The labels and other extraneous matter applied to containers are removed with caustic (which also sterilizes the material) and, the plastic is partially broken down to oligomers. This eliminates places where the polymer was hydrolyzed during service. Then the oligomers can be rejoined with ethylene glycol and/or converted into products with shorter polymer requirements (e.g., plastic bottles converted to fiber for insulation).

7.6 Polyolefins

Neoprene

The idea of making a plastic that could be formed into a fiber was hotly debated in the 1920s. Attempts to make synthetic rubber from isoprene were not successful and it was not obvious to most chemists how linear polymers could be made without the influence of biological systems. Arthur Nieuwland (1878–1936) a chemist and Catholic priest, began work with acetylene and found that he could polymerize it into divinylacetylene. Divinylacetylene further polymerizes to a rubber-like substance when exposed to sulfur dichloride. Dupont Chemicals acquired the patent rights and began research. At Dupont, Arnold Collins (1899-1982) reacted vinylacetylene with HCl to produce chloroprene (2-chloro-1,3-butadiene, analogous to isoprene) in 1930. This compound was subsequently polymerized to produce a compound called neoprene, which is a chemical-resistant rubber.

Fluorocarbons

In the 1920s, the abundance of ammonia (from the Haber-Bosch process) started the expansion of refrigeration applications. But ammonia was very toxic; and introduction of ammonia in pressurized systems into every home was not a reasonable thing to do. At this time, homes had "ice boxes" (insulated cabinets) that were provided with ice delivered door-to-door.

Girls delivering ice (1918)

Source: https://i0.wp.com/www.historybyzim.com/wp-content/uploads/2014/12/2_GirlsDeliverIce.jpg

General Motors and Dupont teamed together to produce fluorochlorocarbons (e.g., dichlorotetrafluoroethane), which had appropriate volatility to be the working gas in refrigeration and was nontoxic and nonflammable. It worked great and made home refrigerators and air conditioning feasible, but there is a problem that will be discussed in Part III.

In the 1930s the issue for Dupont was the fact that the Frigidaire division of General Motors controlled all the patents for these compounds. Roy J. Plunkett (1910–1994) was hired by Dupont and assigned to find a way to get around the Frigidaire patents.

He made tetrafluoroethylene ($F_2C=CF_2$)[134]:

$$CHCl_3 + 2\,HF \rightarrow CHClF_2 + 2\,HCl$$

$$2\,CHClF_2 \rightarrow C_2F_4 + 2\,HCl$$

And stored it in metal cans on dry ice as a liquid (b.p. -76°C). On April 6, 1938, he tried to release some of the gas from a can, but nothing would come out of the can. Even at room temperature the mass appeared to be stuck in the bottom of the can. Plunkett carefully cut the can open and discovered a slippery white material, which proved to be a polymer of tetrafluoroethene, which we now know as Teflon™.

---

[134] Probably via difluorocarbene (CF_2), which would account for the facile polymerization.

Metal Hydrides

Metal hydrides became an interesting area of research during the 1940s as potential high-energy solid rocket propellants. Herbert C. Brown (1912–2004) studying under Hermann I. Schlesinger (1882–1960) at the University of Chicago made sodium borohydride ($NaBH_4$) in 1940. Brown was more interested in the reactions of this compound with organic substrates and left to Wayne University and then to Perdue University (1947) where he did work which wone him the Nobel Prize in 1979. Schlesinger continued with metal hydrides and invented lithium aluminum hydride ($LiAlH_4$) in 1947. The important feature of these compounds is that they readily dissolve even in polar organic solvents because of the low lattice energy of the combination of a very large anion and a very small cation (the anions virtually come into contact). And in solution, the cations are highly solvated.

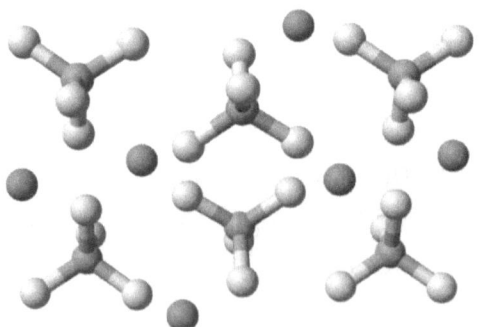

Author: Ben Mills; Source: Wikimedia Commons

Ziegler-Natta Catalysts

Into the 1950s there was still no routine way to polymerize olefins, which are easily isolated from petroleum refining (e.g., ethene, propene, butene). But the work of Schlesinger and Brown was bringing attention of hydride and alkyl compounds of aluminum, which are inherently electron-deficient.

After World War II, Karl Zeigler (1898–1973) followed up the idea of polymerization of olefins (see above discussion concerning manufacture of isooctane fuels) with the objective of forming longer linear polymers. Rather than using Bronsted acids (e.g., HF), he chose a Lewis acid, in particular triethyl aluminum. Triethyl aluminum (and many other metal alkyls such as $LiCH_3$ and $Mg(CH_3)_2$ form "alkyl-bridged" polymers) with electron deficient (three- or four-center, two-electron) bonds:

Author: Smokefoot; Source: Wikimedia Commons

The same thing can happen with hydride (as in diborane, B_2H_6).

Using this technique, Ziegler and Hans-Georg Gellert managed to build polymers with ethylene to about 100 monomer units.[135] These products (trialkyl aluminum compounds) were useful because they decomposed to the 1-olefins and aluminum hydride at high temperature. The terminal olefins could be used to alkylate benzene to make biodegradable detergents (e.g., C_{18}-alkylbenzenesulfonic acid salts).[136]

Note that linear alkyls are readily biodegraded, but branched alkanes are not easily biodegraded. Thus, in the early 1950s there was a problem with foaming in waters receiving (propylene-tetramer)-benzenesulfonates used as detergents, which had been introduced in the 1930s and became widely used after WWII.

In the 1930s, Ziegler had attempted to sublime lithium alkyls, but found that they decomposed to lithium hydride. Attempts to reverse this reaction, e.g., by adding lithium hydride to olefins under pressure were doomed to failure by the high lattice energy of lithium hydride. However, once Schlesinger *et al.* produced lithium aluminum hydride in 1947 and found it was soluble in ether, the problem of lattice energy was resolved. Thus, in 1948 Ziegler and Gellert found that lithium aluminum hydride reacted

---

[135] K. Ziegler and H.-G. Gellert, *Ann. Chem.*, 567:195 (1950).

[136] Previously branched chain alkylbenzenes had been used and these are not readily biodegraded. Thus, the detergents produced foam in industrial waste water released into streams.

smoothly with ethylene at moderate temperature. At 100°C (1 atm), lithium tetraethylaluminate could be isolated; but at 180-200°C and 100 atm long polymers of ethylene are formed.[137]

It was found that direct reaction of aluminum with hydrogen was not effective; but with H_2 and an olefin, the corresponding aluminum alkyls are formed. It was also observed that in some cases, the addition reaction was prematurely terminated. Examination of the materials indicated that traces of nickel in the aluminum produced the effect (Raney nickel). While trying to eliminate this effect, Ziegler et al. found that some transition metals ($TiCl_4$) delayed the elimination of the terminal olefin allowing truly long polymer chains to form. Thus, mixed-metal catalysts were invented.

Giulio Natta (1903–1979) an organic chemist in Italy who (like-Ziegler) rode out WWII in an Axis state by staying out of politics and focusing on research, became familiar with Ziegler's work in 1952. By 1954, he was aware that the poly-olefin polymers occurred in different isomeric forms (isotactic, syndiotactic, and atactic)[138] with different physical properties. These structures

---

[137] K. Ziegler and H.-G. Gellert. *Brennstoffchemie*. 33:193-200 (1952).

[138] "Tacticity" relates to the relative orientation of adjacent centers of optical activity (carbons with four different substituents) within a linear polymer. If all the centers have the same orientation, the polymers fit together more neatly that if they have different (random) orientations. The physical properties of substituted poly-ethylenes depend upon these effects.

were confirmed by x-ray crystallography.   Natta's work not only helped understand the physical properties of polyolefin compounds, it assisted in understanding the mechanism of the polymerization reaction. He received the Nobel Prize with Ziegler in 1963.

www.psrc.usm.edu/macrog/

Author: Stuart Prescott and Robert G. Gilbert; Source:
http://discovery.kcpc.usyd.edu.au/9.2.1short/9.2.1_Polyethene2.html

With Ziegler-Natta catalysts, polyolefins (polyethylene, polypropylene, polybutylene, polyvinyl chloride, polyvinylidene chloride etc. were prepared). These are thermoplastics (most of them melt without decomposition when heated) and can be formed into many shapes by extrusion, and molding.

Polyvinyl chloride (PVC) is extensively use in a variety of applications where rigidity is required including plastic pipes and sheets. PVC is hard to ignite, but when PVC is excessively heated or burned, it generates HCl and benzene, which burns with a sooty flame.

Part III. Chemicals and Life

8.0 Bioactive Molecules

Toxicology and Pharmacology

The late 1800s and early 1900s saw the advancement of toxicology and pharmacology as sciences. We will soon meet Paul Ehrlich who was one of the pioneers in the field of chemotherapy. As Paracelsus (1493-1541) taught *"sola dosis facit venenum."* Obviously, people had been taking potions and herbal remedies since antiquity. The recognition that many human diseases were caused by microbial parasites initiated the idea that chemical compounds (which could be artificial synthesized) added a hypothesis of "relative toxicity" to medicine; and it ushered in an era of scientific study of toxicity and the related science of pharmacology.

You may recall that Perkin (1856) had been making naïve attempts to synthesize quinine when he discovered a purple dye and founded the commercial dye industry. From that success and the realization that diazonium reagents could be used to synthesize series of related compounds with related structures, but different chemical properties, organic chemistry was in a position to make systematic modifications in candidate compounds that might have toxic effects on various organisms.

Thus, the idea of finding drugs that would have toxicity towards to microorganisms but no toxicity (or less toxicity) to humans was born.

Conceptually, within a population of humans or microbes there are some individuals with more resistance to toxic substances than others. The population may fall into a "bell-shaped" curve.

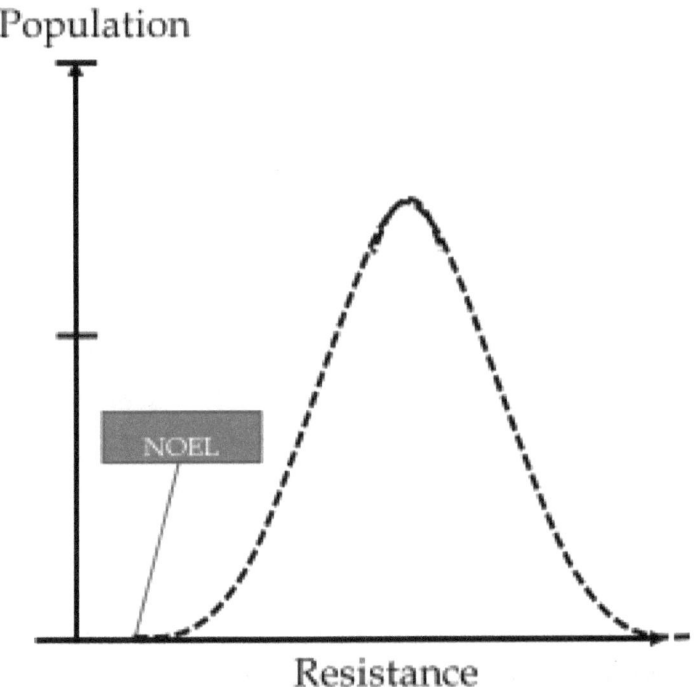

The *No Observed Effects Level* (NOEL) is the resistance shown by the least resistant individuals in the population. As a result, an S-shaped Dose-Response curve is expected. Dose is generally measured in mass of drug per mass of the organism (mg/kg) and may be presented as "dose rate" (mg/kg/day) for substances that may offer chronic or cumulative effects.

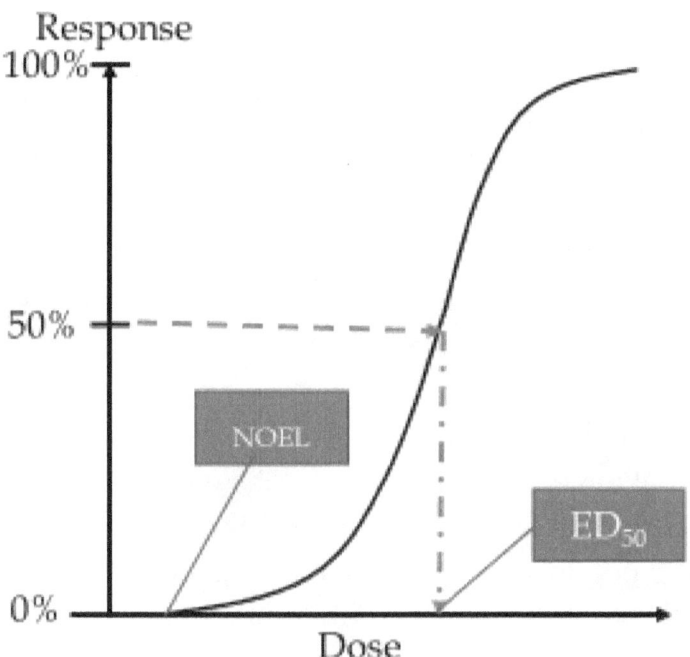

The Effective Dose for 50% of the population (ED_{50}) is the quantity most often tabulated. The <u>Effect</u> may be any observable "biological end-point" such as death. In this case, it is identified as the lethal dose (LD_{50}). These curves are typically used to project risk of adverse effects for an individual. In particular, regulatory standards for exposure are generally set below the NOEL typically with a safety factor of 10 or more. For example, if the NOEL is *10 mg/kg*, the regulatory standards may be set at "*1 mg/kg*."

8.1 The Microbial Theory of Disease

Louis Pasture

Louis Pasteur (1822 – 1895) was a chemist who had done important work demonstrating that the three-dimensional structure of molecules determined their optical activity (1848). However, he soon became interested in the concept of spontaneous generation (the idea that biological organisms spontaneously arise in certain types of environment). This concept (abiotic synthesis) was in some ways tied to the hypotheses of evolution that were becoming more relevant with the work of Charles Darwin (1859) and the abandonment of the theory of vital force[139] (1830s). Because he was (by this time) a leading French scientist, his work that demonstrated that microbes did not develop in sterilized containers won the day and led to the application of pasteurization of milk and beer (1865).

In the 1870s, Pasture became interested in various diseases of commercially important animals (silkworms, chickens). He soon showed that these diseases could be transmitted through cultures of the microorganisms. A series of observations in 1879-80 in which accidentally weakened cultures (e.g., aged) of the microorganism failed to induce lethal infections, he discovered

---

[139] The idea that certain molecules could only be formed by biological systems.

that the exposure of the animals (chickens, sheep) to the weakened culture protected them from clearly potentially lethal doses of fresh cultures. These observations helped understand and generalize the idea of vaccination, which had been practiced using cowpox to mitigate small pox since 1790 (Edward Jenner). Moreover, the idea that a natural disease could be artificially weakened and used as a vaccine was new (but not unique to Pasture as other people were finding similar results in the same period).[140]

Antitoxins

Some bacteria cause disease because they secrete toxins that disrupt the normal functions of cells. Examples include diphtheria, tetanus, botulinum, and cholera. It had been known for a long time that exposure to small doses of toxins (e.g., arsenic trioxide) built some resistance in humans and animals. Thus, a logical strategy in the 1800s was to expose domestic animals (horses) to small doses of toxins and isolate the plasma containing the antitoxin from the animal's blood. This plasma could be concentrated and used as a serum in humans. In this way a large dose of antitoxin could counter the effect of the toxin while the immune system of the infected person defeated the actual infection.

---

[140] Pasture was not a physician and has been criticized for some of his human experimentation, which would now be regarded as unethical. He also obscured some of his findings. Fortunately, his instincts were better than his experimental design and technique.

While diphtheria is very rare in modern countries today, it was a serious killer of children in the 1800s.[141] The toxin was discovered in 1888; and in 1890, Emil von Behring (1854–1917) and Paul Ehrlich (1854–1915)[142] developed an antitoxin serum by inoculating horses with low doses.[143]

8.2 Chemotherapy against Protozoa

Microscopy

The invention of single lens microscopes opened a world of biology never before seen. The largest and first-recognized microorganisms were the protozoa. These are large single-cell organisms that have been evolving for billions of years. Like the cells of higher (multicellular) organisms the cells have nuclei with linear DNA that is encapsulated in a nucleus to protect it from the highly damaging reactive oxygen species (ROS) that leak out of the mitochondria where energy is generated. Because

---

[141] The lethal dose of toxin in humans is 0.1 µg of toxin per kg of body weight. The toxin is a protein that disrupts a human enzyme essential for protein synthesis.

[142] Behring received both the financial benefit and academic credit.

[143] The novel *Arrowsmith* by Sinclair Lewis (1925) follows a physician through his career and covered the period 1900-1930 in very realistic fashion. I recommend this book for all students entering the medical professions.

of their biochemical similarity to human cells these organisms are susceptible to many of the same toxins that are toxic to humans.

Bacteria are much small than protozoa (indeed, many protozoa feed on bacteria). This makes them harder to see and study. Because the metabolism of bacteria is typically much simpler and distinctly different from that of higher organisms, it is less obvious what chemicals might be toxic to them.

The invention of multiple-lens (achromatic) microscopes (1830s) by Joseph J. Lister (1786–1869) allowed study of the organelles that make up microorganisms. Staining these organelles with dyes made many structures visible that had not previously been observed.

Paul Ehrlich

Paul Ehrlich (1854–1915) developed interest in dyes and their ability to stain cells so that they revealed organelles (1882). In 1889 he began work with the dye "methylene blue," which both stained microorganisms and appeared to cure malaria. But methylene blue had adverse side effects (it also stained nerve cells).

Methylene Blue

Author: Harbinary; Source: Wikimedia Commons

He joined Robert Koch's (1843-1910) team in 1891 and gradually developed a receptor theory of immunology and toxicity.

Ehrlich began the search for chemicals that had unique (specific) activity against microorganisms, i.e., a systematic search for chemotherapeutic agents. As a model target organism, he adopted Trypanosomes (a family of protozoa that causes various diseases including African sleeping sickness[144]). He used the guinea pig as his host organism, and soon found that trypan red would kill the protozoa.

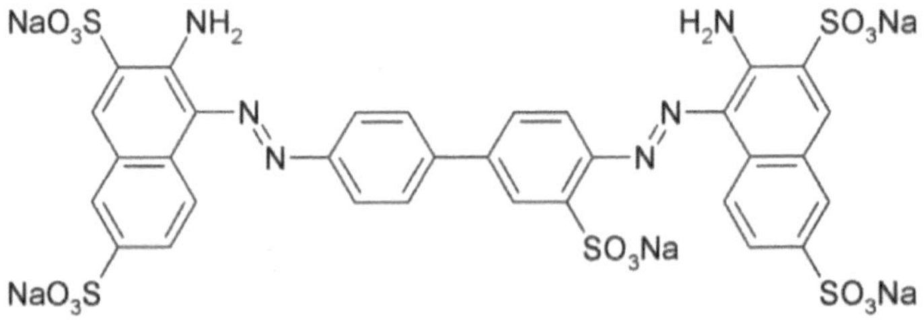

Trypan Red

Author: Klaus Hoffmeier; Source: Wikimedia Commons

Work on dye-related molecules continued in Europe through the 1930s as we will see below. One of the last developments was

---

[144] The cause of African sleeping sickness was identified by David Bruce in 1903. The Tsetse fly is the vector. Koch R. 1904. Remarks on Trypanosome Diseases. *Br. Med J.* 2(2291):1445-9.

"Bayer 205" a sodium salt of a complicated molecule called suramin by the Germans:

Suramin

Author: Fvasconcellos; Source: Wikipedia Media

Suramin was discovered to be effective against trypanosomes in 1917. It was during this time that Ehrlich (in cooperation with Koch) began the systematic screening of various chemicals and developed the dose-response concept for comparing potential drugs.

Colonialism and Medicine

It should be noted that all the European nations were eager to establish colonies in parts of the world considered "backward," mainly Africa and southeast Asia, where society was based on tribal relationships or petty-kings. Central Africa will play a large role in the story. Of course, the Europeans and New Englanders had taken advantage of the tribal conflict in Africa since the 1600s to buy slaves for export to the West Indies and Americas. This practice was stopped officially in 1807 by international treaty. Whereas the demand for slaves in mainland North America had always been limited because the conditions of slavery there actually produced an excess of unskilled labor (principally for agriculture in the southern states)[145]; the sugar business that was practiced in the West Indies (e.g., Hispaniola, Cuba, Jamaica) and the rubber business in Brazil literally consumed slaves as raw material (the typical life-time of a slave in the sugar and rubber industries was 3-5 years of constant labor exposed to malaria and yellow fever). Thus, well after 1807, New England (i.e., Massachusetts, Rhode Island, Connecticut and New York) slave traders were illegally bringing slaves to Spanish, French and Portuguese colonies in the West Indies and South America to feed a demand for expendable labor at least until 1861.

Thus, when France, Germany, Portugal, Britain and the Netherlands carved up central Africa in the late 1800s, the

---

[145] The southern British colonies and later States raised seasonal crops (tobacco and cotton) in an environment largely free from tropical diseases. Thus, slaves became reproducing families in the North American colonies and states: After importation of about 500,000 (in total) slaves in to the mainland British colonies, the population of slaves in the United States grew to around **4 million** in about 70 years.

In this same time period, approximately **12 million** (mainly male) slaves were worked to death in the sugar and rubber plantations of the West Indies and South America: In 1790, the French had about estimated at 450,000 (half of the slaves in the West Indies). When the British freed their slaves in 1833, they only had 800,000. These small numbers suggest the conditions that the 10,000,000 Africans imported to the West Indies and South America (not including Brazil) were very harsh.

essential effect was to enslave Africans in their own country. The Belgians (specifically King Leopold I) took the lead in the central African region, which can be identified as the Congo River basin (watershed).

Few people seem to appreciate the importance of topography in central Africa. The Congo basin is on a high plateau. Winds from the oceans drop an enormous amount of rain here and it is trapped except for a single steep spillway that empties to the Atlantic Ocean. Thus, the rivers on the plateau are broad, slow-moving and meandering. The entire area is swampy and tropical jungle: Perfect breeding grounds for mosquitoes in the forest and testes flies in the fields.

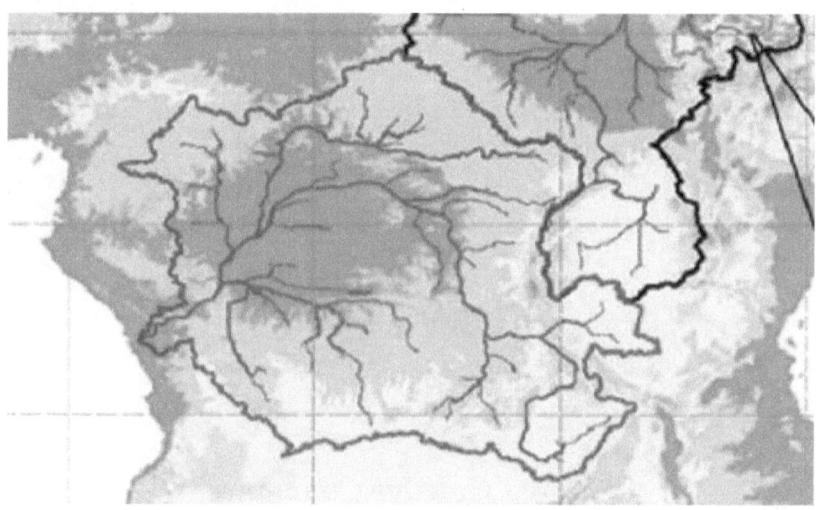

The Congo Basis surrounded by Mountains and Highlands

Author: Mohamed S. Siam; Source: Research Gate

African Colonies after
the Berlin Conference of 1884

Source: Creative Commons

Enormous ecological and evolutionary pressure is placed on humans moving west into the Congo Basin from the plains of the great riff valley. Even today, approximately 30% of children born in this region die from malaria alone before they reach puberty. In this situation any mutation that helps the survival of children (even if it is adverse, to adults) will be incorporated into the population. The result is a series of overall harmful mutations (e.g., sickle cell disease) that allow children to reach reproductive age.

Because the flow of the Congo River is blocked at the western ridge there is a large lake (Lake Kinshasa) and from there the lake overflows a spillway down a long series of rapids and water falls to the ocean. Unlike the Amazon, Mississippi, Mekong, Rhein, etc., oceangoing ships cannot travel inland up the Congo. To exploit their holdings south of the Congo River, the Belgians built (1890-98) a narrow-gauge railroad from Matadi on the coast to Kinshasa (formerly Leopoldville) on the lake. Sections of boats were hauled up to the lake and assembled; and formed the backbone of a river commerce primarily on the Sangha River and upper Congo River. Thus, all products of the Congo Basin in the 1800s and 1900s ultimately came to Leopoldville and were carried to the sea on the railroad controlled by King Leopold I.

The mortality of Europeans in all tropical "possessions"[146] (especially central Africa) was appalling to career bureaucrats.

---

[146] It is hard to call these areas "colonies" they were not being *colonized* in the way that North America was colonized. They were being held and *exploited* for the benefit of the controlling country.

Thus, there was a major interest in defeating tropical diseases and the Belgians created a medical clinic at Leopoldville as early as 1905. By the 1930s the clinic was well established and had conducted many studies (clinical trials of various rigor) on candidate drugs for malaria and sleeping sickness. These trials were mainly conducted on native Africans who coincidentally benefitted in some cases.

Arsenic Drugs

The Germans (led by von Behring, Ehrlich and Koch) dominated medicinal chemistry in the late 1800s and were focused on drugs related to dyes as shown above.

The British African explorer Dr. David Livingstone (1813–1873)[147] and others had reported anecdotally that the arsenic (probably inorganic arsenic As_2O_3) had transient beneficial effects on trypanosome diseases. Thus, it is not surprising that arsenic compounds were screened for their effects against microorganisms. The initial work focused on atoxyl[148] (so named

---

[147] Livingston was a Christian missionary who roamed over much of Africa with his primary objective to end the Portuguese and Arab slave trades of the mid- and late 1800s.

[148] Atoxyl had been synthesized by reaction of arsenic acid (H_3AsO_4) with aniline by Antoine Béchamp in 1863. He apparently though it was a phenylamide of arsenic acid or an anilinium salt. Thus, into 1907, physicians assumed that it was destroyed by acid and refrained from oral administration.

because it is less toxic to humans that arsenic trioxide). In Britain, H. W. Thomas and A. Breinl found that atoxyl killed trypanosomes in animal tests (1905).[149] By this time (1907), it was recognized that the parasites crossed from the blood into the cerebral spinal fluids through the "blood-Brain Barrier" (fatty tissue) where they could not be reached by water-soluble salts.

Robert Koch led an expedition into Africa (Tanzania, 1907) and found that about 2% of patients treated with atoxyl were blinded by the drug. Meanwhile, in Germany, Alfred Bertheim (a chemist working with Ehrlich) determined the correct structure of atoxyl which facilitated making a variety of derivatives of the organic compound including azobenzene derivatives and what were believed to be the analogous arsenobenzene derivatives.

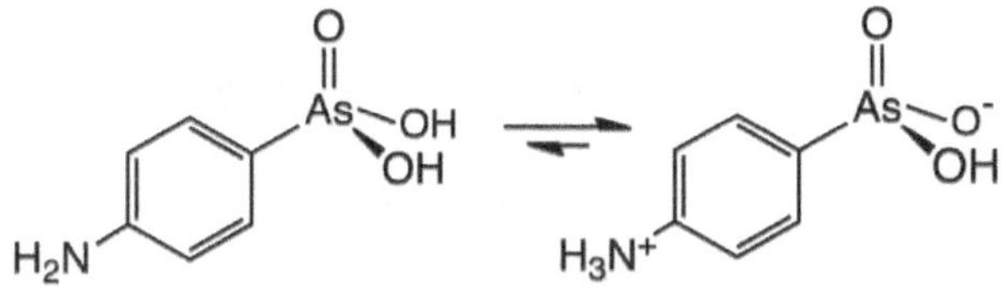

Atoxyl (Arsanilic acid)

Author: Smokefoot; Source: Wikimedia Commons

These compounds were not effective against trypanosomes; but in 1909, Ehrlich's bacteriological associate Sahachiro Hata (1873–

---

[149] Breinl A, Todd JL. Atoxyl in the treatment of Tryoanosomiasis. *British Med J.* 1907 1(2403):132-4.

1938) found that their compound number "606" was effective against the spirochete that causes syphilis. The compound was identified as arsphenamine and given the tradename Salvarsan. It is now known that it is not an arseno-benzene (-As=As-), but rather a family of compounds with rings of arsenic atoms:

Author: Benjah-bmm27; Source: Wikimedia Commons

Trypanosomiasis (African Sleeping Sickness)

One of the results of World War I (1914-18) was redrawing the map of colonial Africa (as well as the Middle East). For the most part, France and Britain acquired the German colonies. Interestingly, a drug that worked reasonably well against sleeping sickness had been developed in Germany in 1917; and in 1922, the German government offered to reveal the formula to the British and French for the return of the former German possessions in Africa. However, any bargaining power the

Germans[150] may have had was undermined by the workers at the Rockefeller Institute in New York.

As usual, a team of a chemist Michael Heidelberger (1888–1991) and a physician Louise Pearce (1885–1959)[151] was working to synthesize derivatives of atoxyl and test them against trypanosomes at the Rockefeller Institute. They discovered that compound "A63" had activity against the trypanosomes that cause African sleeping sickness (1919) *in vitro*. But there were no human subjects to test the drug on in New York City. Thus, they waited for the report of an outbreak of sleeping sickness in Africa.

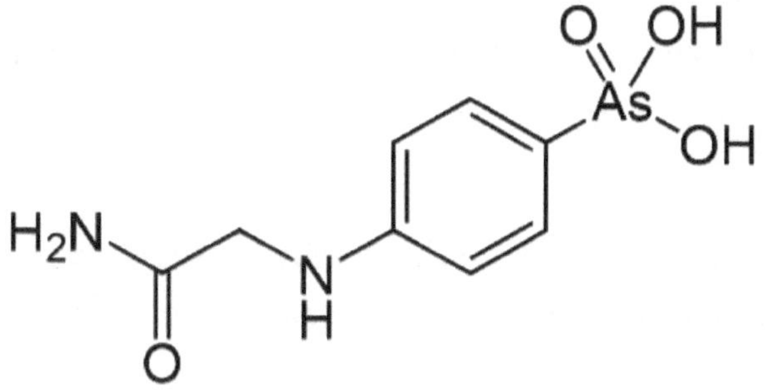

A63/Tryparsamide

Author: Yikrazuul ; Source: Wikimedia Commons

---

[150] The hubris (racism and arrogance) of Europeans regarding Africa in the 1920s when black Africans had been voting and holding office in the United States for over half a century is remarkable.

[151] Notable as one of the first female MD, PhD scientists.

Unlike malaria, sleeping sickness is episodic and seems to appear at random with devastating effect on people of all ages. It is carried by the tsetse fly and initially infects the blood with limited effect. But it soon moves to the brain and slowly degrades the motor skills and awareness of victims who typically waste away over a period of time.

Word arrived in New York that sleeping sickness patients were being treated in substantial numbers in Leopoldville in 1920. Thus, Louise Pearce went to Brussels and obtained permission to travel to the Belgian clinic in Leopoldville and conduct human trials. From Brussels she travelled by steamer to Matadi and then by train to the clinic on the shores of Lake Kinshasa. She wrote the following letter to Heidelberger upon her arrival:

17 July 1920

…

"The director [Dr. Par F. Van den Branden] *has charge of the negro hospital in which there are about 70 advanced cases of sleeping sickness undergoing the routine atoxyl & emetic treatment – 0.5 g atoxyl on Monday – 0.1 emetic on Thursday. I was given 3 relapses to begin on – a most unfortunate type of case – but the next day, an Early untreated patient arrived. 2.0 grams of [tryparsamide] cleared the cervical lymph glands of trypanosomes within 21 hours and I am now waiting to see when a relapse takes place. I hope to get more similar cases from* **the nearest epidemic center – about 500 miles away!** "

It turned out that most of the patients being treated in Leopoldville were actually arriving by boat from Cameroon via the Sangha River a major tributary of the Congo. She was in the Congo for about 6 months and returned to New York and published her favorable results: [152]

Fig. 3. — Laboratoire de Léopoldville. — Salle des cours pratiques.

Source: https://kosubaawate.blogspot.com/2012/11/leopoldville-1902-first-hospital-for_26.html

---

[152] Pearce L. 1921. Studies on the treatment of human trypanosomiasis with Tryparsamide (the sodium salt of N-phenylglycineamide-p-arsonic acid). *Journal of Experimental Medicine.* xxxiv(6) supplement 1: 1-104.

The object of the mission sent to the Belgian Congo was the obtaining of concrete facts and observations as outlined above. All of the work was done at the government laboratory and Hôpital de la Reine in Léopoldville. The helpful cooperation of the Belgian and Colonial Governments made it possible to accomplish a considerable amount of work in a comparatively short time and it is a pleasure to express here our appreciation of the aid so cordially extended to us, first in Brussels and then in the Congo. We also wish to acknowledge the assistance given us by Professor Rodhain, Chief of the Colonial Medical Service, and by Dr. van den Branden, Director, and Dr. van Hoof, Assistant Director of the Laboratory in Léopoldville.

SOURCE AND TYPE OF PATIENTS.

The observations which form the basis of this report were made on 77 native cases of trypanosomiasis caused by *Tr. gambiense* which were treated with tryparsamide in Léopoldville, Belgian Congo. The

civil service of the Belgian Congo has established throughout the colony, medical stations and posts under the charge of government physicians and traveling sanitary agents, in addition to which a considerable amount of medical work is done by various religious missions. The native population is thus examined for trypanosomiasis at more or less regular intervals, the routine procedure consisting of the palpation of the cervical lymph glands and the microscopic examination of lymph gland juice from those individuals having one or more palpable cervical glands. At the present time, there is no widespread epidemic of a severe character in the immediate vicinity of Léopoldville but there are many endemic foci of the disease in this district, while a few days' travel brings one into active epidemic areas. The patients treated with tryparsamide were entirely typical of the routine run of trypanosomiasis cases met with in Léopoldville. The population of the nearby native villages comes *en masse* to the laboratory every 3 months for examination and several cases were obtained in this way.

Moreover, Léopoldville is the terminus of one of the large river transportation companies and a medical passport issued after examination for trypanosomiasis, is required of any native traveling from one part of the colony to another. The entire crew of each boat is examined the day before sailing and this source yielded a number of cases from distant parts of the colony. Finally, several advanced cases under treatment in the native Hôpital de la Reine and lazaret in Léopoldville were transferred to us for treatment with tryparsamide. All of the patients studied thus fell into three general classes: First, those who were sent to the lazaret, which is in reality a native village under nominal native police control on the outskirts of Léopoldville, and who came to the laboratory when sent for; second, patients in the Hôpital de la Reine, over whom a closer supervision could be maintained; and third, ambulatory cases, corresponding to dispensary patients, who continued to live in their own villages and who came to the laboratory at certain fixed times.

Source: Downloaded from www.jem.org

Based on this success, Tryparsamide became the standard treatment for sleeping sickness and was used by Eugène Jamot (1879–1937) to virtually eliminate sleeping sickness from the Congo Basin between 1925 and 1935. However, sleeping sickness periodically returns when surveillance is not aggressively pursued.

Treatment of Malaria

Malaria is also caused by a protozoan, but the lifecycle of the protozoa (Plasmodium) in humans is more complex that trypanosomes. Like butterflies, plasmodia go through a series of fundamental changes in its complex life cycle. It is harder to study and harder to treat. The insect vector is the mosquito, which punctures the skin and injects anticoagulant into the victim's blood. The first stage of the plasmosium life cycles in humans is as sporozoites that enter and takeover liver cells by suppressing apoptosis (i.e., cell suicide). In these cells the parasite changes form and produces many schizontcites, which break out and infect the erythrocytes (red blood cells).

It appears that malaria was either not present (*Plasmodium falciparum*) or only present in relatively mild forms (*P. vivax*) in the western hemisphere (the Americas) before the Europeans and Africans arrived.[153] Ironically, a folk medicine used by the native South Americans obtained from the bark of the quina-quina trees (Europeans call them cinchona) suppresses the disease and was adopted by Europeans as a remedy for malaria as early as 1630. In the 1800s, chemical techniques had advanced to the point where quinine was extracted from the bark and purified for use.

The British in particular ritualized the use of quinine as "tonic water" and mitigated the bitter taste by diluting the extract with gin (an alcoholic drink flavored with juniper berries). Thus, the "gin and tonic" protected many British colonial governors in Africa and Asia for three hundred years with no idea how the drug worked.

---

[153] Rodrigues, PT et al. Human migration and the spread of malaria parasites to the New World. *Sci Rep.* 2018. 8(1):1993.

Quinine

Author: Vaccinationist; Source: Wikimedia Commons

Even today the mode of action is not entirely clear. It is generally accepted that quinine does not have any effect on the parasite until it reaches the erythrocytes. As a mildly basic compound, quinine is concentrated in the lysosomes of the malaria parasite and interferes with its metabolism. The parasite ultimately invades red blood cells and consumes hemoglobin. This naturally causes great problems for the host because that starves the cells for oxygen. In the parasite, the hemoglobin is converted to heme. The heme would be toxic to the parasite, but is polymerized into a form (hemazoin) that is not toxic. Apparently, quinine (and similar drugs) interferes with this last step and the build-up of heme kills the parasite. Overall, quinine is known to interfere with nucleic acid synthesis, protein synthesis, and glycolysis.

By 1924, the empirical formula (i.e., ratio of C:H:N:O) of quinine was well known and some features of the structure had been deduced. In Germany chemists Warner Schulemann, Fritz Schoenhoeffer and August Wingler managed to synthesize a compound that had these features called "plasmochin" in Germany and now "pamaquine."

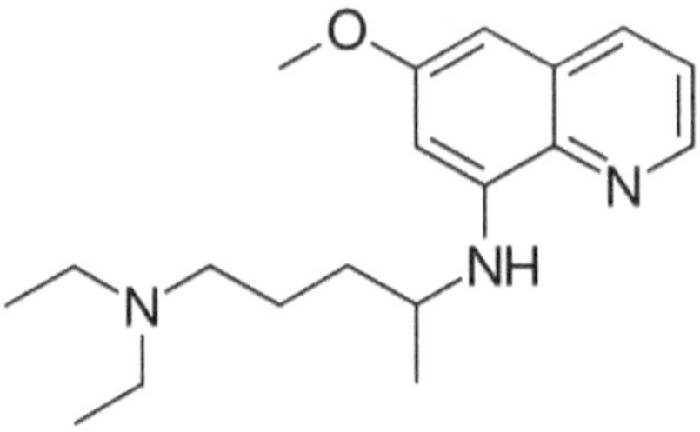

Pamaquine 1927-1942

Author: Fvasconcellos; Source: Wikimedia Commons

After finding that the drug had some favorable effects on malaria in canaries, a human trial was conducted in early 1927 at the Belgian clinic in Leopoldville. The drug was effective but had side effects on various cellular blood components. Nonetheless, it was used into the 1940s. In 1934, resochin (chloroquine) and sontochin ("3-methyl-chloroquine") were synthesized by Johann Andersag (1902-1955) while working for the Bayer Company.

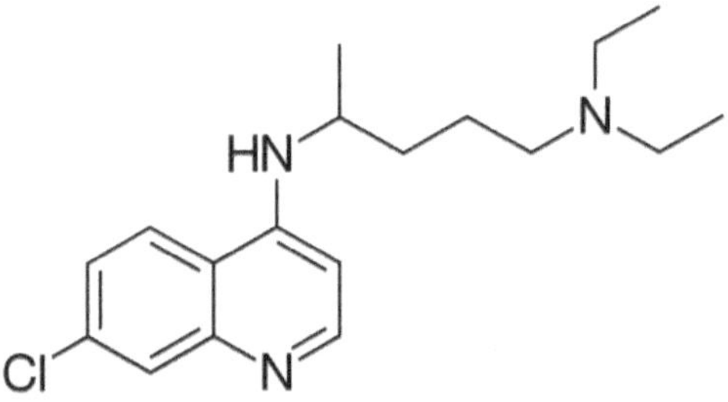

Chloroquine

Author: Fvasconcellos; Source: Wikimedia Commons

He was obviously working under the supervision of Warner Schulemann, whose stamp appears at the bottom of the page of the Bayer laboratory notebook shown below.

Author Johann Andersag; Source Bayer/Wikimedia Commons

But Schulemann[154] did not mention this work when he was interview by American intelligence officers after WWII (May 1945). Although it is generally reported that these compounds were viewed as too toxic to use, it is clear that the Germans did in fact issue sontochin to their troops in North Africa.

---

[154] American intelligence officers interviewed, and in some cases incarcerated, German and Japanese scientists after the war.
https://collections.nlm.nih.gov/ext/dw/101641398/PDF/101641398.pdf

Meanwhile, the formula for resochin (chloroquine) was passed to an American company during the 1930s. When French troops captured German supplies in North Africa (~1942) and forwarded sontochin ("3-methyl-chloroquine") to the Americans, the American research program figured out that chloroquine was the drug of choice. It proved to be very effective and produced few side effects.

Chloroquine is still used today, but (as will be discussed below) the protozoa have developed substantial resistance and it does not work very well. For this reason, modern research has begun to focus on other (dissimilar) drugs. Another folk remedy is sweet wormwood from China. In 1972, Tu Youyou identified this compound as a potential starting point for antimalarials.

Author: Lukáš Mižoch; Source Wikipedia Commons

8.3 Chemotherapy against Bacteria

In the early 1900s bacteria were found to be a major source of human disease. Although antitoxin strategies and vaccination strategies were available, certain classes of bacterial infections seemed to be beyond the range of chemotherapeutic treatments.

The first viruses were recognized; not because they were seen but because they were "filterable," i.e., they were so small that they passed through filters and still caused diseased.[155] And some large viruses called phages were identified. Phages were interesting because they actually attacked and destroyed bacteria.[156]

Sulfur Drugs/ Prontosil and Sulfanilamide

The ability to convert aromatic amines into diazonium compounds that could be coupled with aromatic rings activated by electron donating amine or phenol substituents yields an almost unlimited number of different compounds to study. In Germany, the Bayer Wuppertal-Elberfeld dye factory (*I.G. Farbenindustrie*) continued research on azo-dyes after WWI. In particular, they had a program in which chemists Josef Klarer and Fritz Mietzsch produced compounds for pharmaceutical

---

[155] Wolbach SB. 1912. The Filterable Viruses, a Summary. *J Med Res.* 27(1):1-25.

[156] Bronfenbrenner JJ, Korb C. Studies on the bacteriophage of D'Herelle: IV. Concerning the one ness of the bacteriophage. *J Exp Med.* 1925. 42(6):821-8.

Gratia A. Studies on the D'Herelle phenomenon. *J Exp Med.* 1921. 34(1):115-26.

evaluation by Gerhard Domagk (1895-1964). In 1931 they produced a dye called "prontosil red."

Prontosil
Author: Edgar181; Source
Wikimedia Commons

This compound is not particularly soluble and was not effective against bacterial *in vitro*. But prontosil could be administered orally to test animals (mice and rabbits) and Dolmagk (1932) found it to be protective *in vivo* against normally lethal injections of gram-positive cocci (staphylococci and hemolytic streptococci) that cause infections of the blood that result in sepsis.[157] Thus, from an early date, Domagk suspected that the drug might be metabolized to an effective antibacterial agent. Possibly because of these observations, he was not initially optimistic that the dye would be effective in humans. His "hand was forced," however, when his daughter developed a serious streptococcal infection. He gave her a dose of prontosil and she soon recovered.[158] This observation motivated him to ask clinicians to evaluate the drug, which resulted in a formal publication "*Ein Beitrag zur Chemotherapie der bakteriellen Infektionen*" (A contribution to the

---

[157] Sepsis is a result of an immune response to a variety of infectious agents that are not in themselves lethal.

[158] About this time Bayer patented prontosil in the name of the chemists who synthesized it (1933).

chemotherapy of bacterial infections), *Deutsche Medizinische Wochenschrift* (1935). The drug was then marketed but was not widely used because bacterial infections were generally assumed to be untreatable by physicians. In 1936, Franklin Delano Roosevelt Jr. (age 22, son of the President of the US) developed a similar infection and was successfully treated with prontosil. This event launched prontosil popularity throughout the United States. Domagk was awarded the Nobel Prize in 1939, but at that time, Germany was at war with Britain and France. Thus, he was not able to receive the prize until 1947.[159]

Meanwhile, other people were interested in the field and probably were looking toward circumventing the Bayer patent on prontosil. Tréfouël et al.[160] discovered that the actual active agent is 4-aminophenylsulfonamide (1935).

Author: Choij; Source: Wikimedia Commons

---

[159] Gerhard Domagk – Facts. NobelPrize.org. Nobel Media AB 2019. Sun. 3 Nov 2019.
https://www.nobelprize.org/prizes/medicine/1939/domagk/facts/

[160] J. et Th. Tréfouël, F. Nitti et D. Bovet, « Activité du p.aminophénylsulfamide sur l'infection streptococcique expérimentale de la souris et du lapin », dans *C. R. Soc. Biol.*, vol. 120, 23 novembre 1935, p. 756.

This compound had actually been synthesized by Paul Gelmo (1879–1961) and was patented in 1909 because it is useful in forming many classes of compounds via the azo coupling reaction (e.g., it was one of the starting materials used to make prontosil). Not only is this compound soluble, it is readily derivatized and quickly gave way to a large family of related "sulfa drugs" with activity against different organisms.

It turns out that the mechanism of action of sulfa drugs is "bacteriostatic," they do not kill the bacteria, they prevent the bacteria from multiplying. Bacteria make folic acid, which is essential for making DNA.

Folic Acid

Author: Calvero; Source: Wikimedia Commons

4-Aminophenylsulfonamide interferes with the synthesis of folic acid (which does not kill the organism, but limits the replication of the bacteria). Higher organisms obtain folic acid as a vitamin (we do not synthesize it); thus, the sulfa drugs are not particularly toxic to humans.

Penicillin

Alexander Fleming (1881– 1955) was a bacteriologist. During WWI, he realized that antiseptic (e.g., phenol applied to open wounds following the experiences of Joseph Lister) often damaged tissue without getting at the infective agents that had been pushed into the wound. So, Fleming was well aware of the horrors of war and short comings military medicine.

In 1928, while studying staphylococci bacteria grown on agar media in petri dishes, he noticed that a mold (which he identified as a Penicillium) had colonized one dish and that bacteria were not able to grow in a significant ring around the mold.

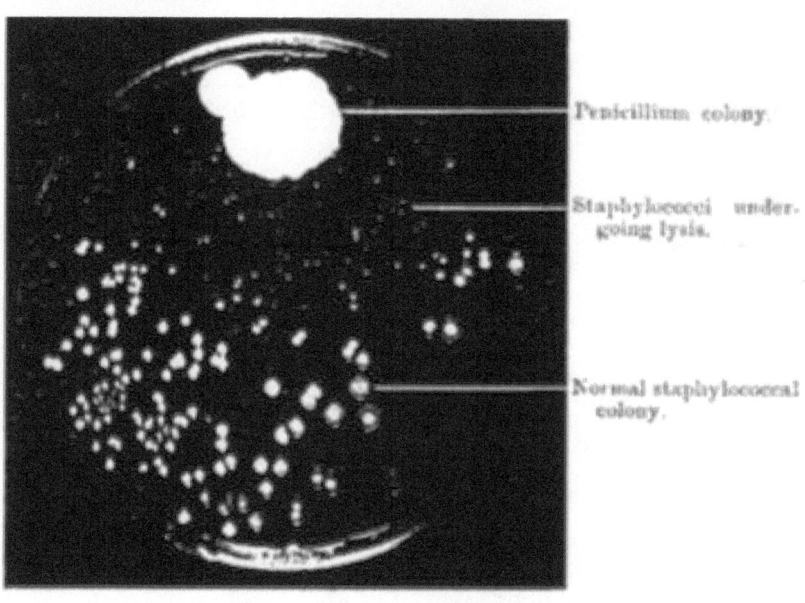

Penicillium colony.

Staphylococci undergoing lysis.

Normal staphylococcal colony.

Figures from Flemings Original Paper

He attempted to cultivate the mold and isolate the liquid that he called penicillin (March 1929) but he was a biologist (not a chemist) and was not able to make any progress.[161] But he did publish his observations.

The really hard work was not undertaken until the beginning of WWII. In 1939, the British government was refocusing all of its energies toward the war with Germany. Only research projects with immediate military relevance were being funded.

---

[161] Bentley R. The development of penicillin: genesis of a famous antibiotic. *Perspect Biol Med*. 2005. 48(3):444-52.

https://www.ncbi.nlm.nih.gov/pmc/articles/PMC2566493/pdf/11545337.pdf

Obviously, an anti-bacterial drug was relevant to the war effort. Howard Florey (1898–1968) and chemist Ernst Boris Chain (1906–1979) put together a team dedicated to producing penicillin in useful quantities. Relying heavily on the skills of chemist Norman Heatley (1911–2004) to develop scalable isolation techniques for the drug that proved to be chemically unstable, the team succeeded in isolating enough of the drug to demonstrate efficacy in mice and humans.

Penicillin actually kills bacteria by inhibiting the enzyme that makes components of the cell membrane. Without these components, the bacterial membrane ruptures (lysis as first noted by Fleming). During the war, the purity of the drug was still in question and its structure was unknown. It was not until the x-ray crystallography work of Dorothy Crowfoot Hodgkin (1910–1994) was complete in 1945 that the structure of the drug was known.

Penicillin

Author: Yikrazuul; Source: Wikimedia Commons

The fused ring structure (beta-lactam) is key to its bactericidal action. A number of derivatives are possible where the "R" group is changed. The great strain in this ring system renders the molecule unstable and that is one of the features that made it hard to isolate.

Drug Resistance in Microorganisms

In the late 1940s and early 1950s, chloroquine and penicillin gave an aura of invincibility to medicine. Coming out of WWII, chemists had apparently created an entirely new and powerful paradigm for medicine. The most threatening disease were conquered and now all we had to do was mop up the rest. But it is never so simple. By the 1960s "anti-biotic resistance" was becoming apparent. For example, bacteria evolved a way to degrade penicillin with an enzyme (penicillinase). And it has become apparent that this and other drug-resistance enzymes are rapidly inherited and even *shared* among microorganisms (even across species lines) on extrachromosomal DNA called plasmids. Bacteria can become armed with resistance to drugs they have never even been exposed to. Today, multi-drug resistant strains of microorganism threaten to undo the progress made in chemotherapy. And a similar problem is developing in their insect vectors.

8.4 Pesticides and Herbicides

Natural Products

Insects (6-legs), arachnids (8-legs) and other "bugs" have long been hated for their damage to crops and general nuisance. By

1900 it was also becoming obvious that mosquitoes and flies (especially biting flies) were spreading disease caused by microbes. Arsenic compounds (Paris Green and lead arsenate) had found use in orchards to prevent damage to grapes, apples, pears and citrus fruit. By the 1930s, a composition of chromium(VI) copper(II) and arsenate (e.g., chromated copper arsenate) was being used to pressure-treat wood to prevent decay and termite attack. This product could make quickly grown pine lumber resist rot better than ancient hard woods; thus, facilitating economical construction of wooden houses, railroad ties, utility poles, pilings and bulkheads. Along the way, arsenic developed a particularly bad and undeserved reputations as an environmental hazard (this will be explained below). The first synthetic organic-based insecticides appeared during World War II.

Pyrethrum and Rotenone

In the early 1900s, the only known organic insecticides were pyrethrum and rotenone.

Rotenone is isolated from a vine that has been used to kill fish in native cultures. As early as 1848, it was used to kill caterpillars on various plants. It was first extracted and isolated in 1895 and has been used to kill or capture fish and kill a variety of insect pests on plants. It is not persistent and is sold today as an "organic" (i.e., natural) pesticide:

Rotenone

Author: Jü; Source Wikimedia Commons

Pyrethrum is produced by a variety of flowering plants and has been used as an insecticide since antiquity. It is also a natural insecticide. However, it can be modified to make semi-synthetic derivatives.

Pyrethrum

Author: CYL; Source: Wikimedia Commons

Obviously, it would be enormously expensive to synthesize either of these compounds commercially.

Organophosphates

This story begins with Wladimir Moschnin (1854) followed by Philippe de Clermont (1831–1921) who made the ester tetraethyl pyrophosphate from silver pyrophosphate and ethyl iodide:

Author: Rifleman 82; Source: Wikimedia Commons

Though early chemists mention the taste and odor of this compound (which is now used as an insecticide) no one was particularly concerned about phosphate esters until the 1930s. The sequence of events is not clear, but Willy Lange and Gerda von Krueger reported as follows in 1932:

> "*Interestingly, we report the strong effect of* **monofluorophosphate phosphoric acid alkyl esters** *on the human organism. The vapor of these compounds have a pleasant odor and sharply aromatic. After only a few minutes of inhaling the vapor, there is a strong pressure on the larynx, associated with shortness of breath. Then comes decreased awareness, opacities, and dazzling phenomena causing painful sensitivity of the eye to light. Only after several hours is there relief from these phenomena. They are apparently not caused by acidic decomposition products of the ester, but they are probably due to*

the dialkyl monofluorophosphates themselves. The effects are exerted by very small amounts."

Quote from Wikipedia[162]

The Nazi government took over Germany in 1933 and secretly began rearmament. Germany had been leaders in use of chemical weapons in WWI (chlorine, phosgene, mustard gas) and it is likely that this research group are I. G. Farben realized that there would be interest in toxic chemical weapons. If they had already made *monofluorophosphate phosphoric acid alkyl esters* by 1932 and knew they had potent toxic effects, it is clear that by the time that Gerhard Schrader (1903–1990) synthesized "Tabun" in 1936 they were well on their way to effective chemical warfare agents. Indeed, a plant to manufacture Tabun as a weapon of war was established at Dyhernfurth (now Brzeg Dolny, Poland) shortly after occupation by the Nazis in 1939. The plant became operational in 1942 and eventually produced 12,500 tons of the material. Fortunately, by the time it became evident that Germany would lose WWII (after the successful invasion of June 6, 1844) the Allies dominated the air and transportation of

---

[162] Petroianu GA. The synthesis of phosphor ethers: who was Franz Anton Voegeli? *Pharmazie.* 2009 Apr;64(4):269-75.

Petroianu GA. Toxicity of phosphor esters: Willy Lange (1900-1976) and Gerda von Krueger (1907-after 1970). *Pharmazie.* 2010 Oct;65(10):776-80.

chemicals as toxic as Tabun would have been extremely dangerous as the Americans were systematically attacking rail and road targets.

Apparently, the work by I.G. Farben during the war was actually focused on making less-toxic materials. This research produced parathion and several related compounds, which became widely used insecticides in the 1950s.

Author: Smokefoot; Source: Wikipedia Commons

DDT

Many diseases cause disability and death among soldiers and civilians alike during war. We have already seen how the allies in WWII developed chloroquine and penicillin in amazingly short periods of time (1939-1943). Similarly, by screening compounds provided by chemists from around the world an amazing pesticide was discovered.[163]

---

[163] DDT-Weapon against disease. War Department Film Bulletin No. 195. 1945.

The USDA laboratory in Orlando, FL was commissioned by the Department of the Army to identify new insecticides by screening candidates provided from world-wide sources. They needed a compound that could be mass produced and effective against flies, ticks, fleas, lice, and mosquitos. On November 16, 1942 a package arrived from Switzerland where a new pesticide had recently been patented (1940) and marketed (1942). Paul Müller (1899–1965), a chemist working in Switzerland for , J.R. Geigy A.G, had discovered that dichlorodiphenyltriclhoroethane (DDT, 1,1,1-trichloro-2,2-bis(4-chlorophenyl)ethane), which had first been synthesized in 1874 by Othmar Zeidler (1850–1911), was quickly lethal to flies and persistent when applied.

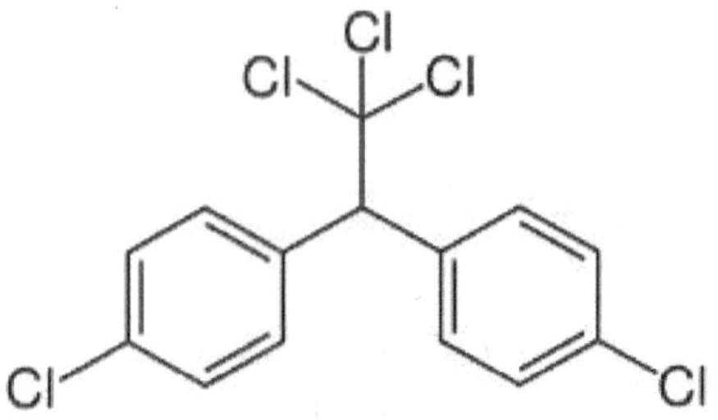

Source: Wikimedia Commons

His decision[164] to make this compound (by the method of Chattaway and Muir in the *J. Chem. Soc.*, 1934, p. 701) was based upon observations by Dr. H. Martin et al. that bis(4-chlorophenyl) compounds of oxygen, sulfur and the sulfoxides

---

[164] Nobel lecture 1948.
https://www.nobelprize.org/uploads/2018/06/muller-lecture.pdf

were toxic when ingested by moths and his own observations that chloroalkyl compounds frequently were toxic to insects. The key to the contact lethality of DDT seemed to be its ready uptake as a non-polar compound (by vapor) by insects that landed on the surface where it was applied. But, the 4,4′-isomer is far more toxic to insects than any combination of other bis-monochlorophenyl isomers. The toxicity to insects seems to be related to the ability to undergo elimination of HCl. For example, the aromatic chlorines can be replaced with fluorine or methoxy groups while retaining the toxicity. The 4,4′-fluro compound is faster acting and the 4,4′-methoxy compound has a narrower range of effectiveness (i.e., more specificity to insect pest).

This photograph appears in many internet sources

https://www.acsh.org/news/2016/02/23/zika-virus-infections-in-the-americas-and-the-ddt-question

There are many photos on the Internet of children playing in the dust of DDT being sprayed to eliminated insect pest (especially flies and mosquitoes). Overall, DDT was so successful and seemed to be so benign towards humans, plants and animals that it became very widely used in the civilian population immediately after WWII.

The use of DDT and chloroquine quickly eliminated the threat of malaria and yellow fever from the southern United States. In the period 1955-60, the World Health Organization (WHO) attempted to eradicate mosquito-borne diseases for the tropics with these chemicals. They succeeded in temperate areas and on small islands in tropical settings, but they were not successful in central Africa and South East Asia. Indeed, by this point, the problem of resistance of both insects to DDT and *Plasmodium* to chloroquine was becoming evident.

In the meantime, a number of chlorine-containing insecticides were introduced including: dieldrin, aldrin, chlordane, lindane, mirex, kepone and toxaphene.

Author: Leyo; Source: Wikimedia Commons

In the 1940s and into the 1950s, it was assumed that most compounds would be destroyed in the environment. Thus, it was common to advertise that these compounds would continue working for many years. No one thought much about this until 1962. To explain this, change in attitude, we need to go back and pick up several topics.

Herbicide 2,4,5-T

It turns out that plants have growth hormones (auxins e.g., indole-3-acetic acid). These hormones have no effect on animals. It was discovered (1940) that high doses of plant auxins can cause plants to "grow themselves to death" while showing virtually no toxicity to humans, mammals, birds, fish or insects. Interest in developing persistent (thus, requiring only one dose) versions of these chemicals during WWII was associated with defoliating jungles and potentially use against food crops.

Indole-3-acetic acid

Author: Ayacop; Source: Wikimedia Commons

2,4,5-T

Author: Monolemma; Source: Wikimedia Commons

The compound 2,4,5-trichlorophenoxy acetic acid (2,4,5-T) and esters, and the related 2,4-D compounds, are more slowly metabolized and are toxic to broad-leaf weeds (which absorb them well and grow quickly).

While they have few objectionable properties in themselves, there were cases where workers manufacturing 2,4,5-trichlorophenol broke out as an acne condition that was not understood. Going into to the Vietnam War (1965-75) American forces used these herbicides (as "agent orange") extensively to eliminate ambush sites along roads and trails. Again, this acne was occasionally observed in production workers.

It was discovered that the acne ("chloracne") was the result of contamination of 2,4,5-T by 2,3,7,8-tetrachlorodibenzodioxin (TCDD, a.k.a., dioxin). TCDD and other isomers and related compounds (dibenzofurans) inadvertently are produced as byproducts when the salt of the 2,4,5-trichlorophenol was reacted with chloroacetic acid:

2,3,7,8-Dibenzodioxin

Author: Emeldir; Source: Wikimedia Commons

The phenol oxide of one molecule displaces the chloride adjacent to the oxygen of another.

TCDD coincidentally fits into an important receptor (aryl hydrocarbon receptor) common in mammalian cells and causes (among other things) excessive excretion of oil via the skin (perhaps as a detoxification mechanism for this molecule that is not very soluble in water and tends to collect in fat). The toxic effects are different in different mammals. In some, the compound is very lethal in others it is not particularly lethal but may cause developmental effects or cancer. After much debate, long-term epidemiological studies from Vietnam and studies on worker receiving high doses of 2,4,5-T and TCDD in occupational settings, it is clear that the risk of dioxin as a carcinogen are low or non-existent.[165]

---

[165] Buffler PA, Ginevan ME, Mandel JS, Watkins DK (September 2011). "The Air Force health study: an epidemiologic retrospective". *Ann Epidemiol*. 21 (9): 673–87. I might mention that about 1960, my father was clearing land behind out house for a field and I was assigned at task of killing brush. I was armed with a hatchet and a leaky oil can filled with an oil contain 2,4,5-T so that I could go from bush to bush make a cut and squirt in the herbicide. As I recall, I worked for days with the oil on my hands, arms and clothes. Yet, here I am in 2020.

Glyphosate and Genetically Modified Crops

Of the twenty-odd amino acids, higher animals have lost the ability to synthesize most of those that have aromatic side chains (i.e., phenylalanine, tyrosine, and tryptophan produced via the shikimate pathway). We are dependent upon food sources for these amino acids, which are used in our enzymes; but we are indifferent to any molecule that merely blocks the key enzymes in the shikimate pathway.

Water that contains calcium and magnesium ions is called hard water, because these ions cause the precipitation of fatty acid anions typically found in soaps as sodium salts. This phenomenon causes the "ring around the bathtub" and uncomfortable deposits on skin and hair. Thus, many people have looked at anions that would chelate (i.e., selectively bind with) calcium and magnesium more strongly than fatty acids and keep them in solution. A typical approach would be to have a carboxylic acid on one end of a molecule (pKa about 2.5, similar to fatty acids)[166] and on the other end a very ionic group like phosphate. Phosphate and sulfate substituents are used to keep organic molecules (e.g., naturally occurring phosphatidylserine glycerides) in solution.

---

[166] The pKa is the pH when the acid is 50% ionized. Thus, it is a measure of acid strength. The lower the number, the stronger the acid.

Source:
https://avantilipids.com/wpcontent/uploads/2015/11/phosph
atidylserine-diagram.gif

Thus, there had been experimentation with many phosphate and phosphonate compounds with similar structure. Indeed, N-(phosphonomethyl)glycine appears to have been synthesized in three different laboratories starting in the 1950s as a potential water softening agent.

Author: Yikrazuul; Source, Wikimedia Commons

However, Monsanto also happened to be in the agricultural chemical business and screened compounds as herbicides and pesticides. This work led John E. Franz (1929-) to synthesize the compound as an herbicide.

It has been determined that N-(phosphonomethyl)glycine binds to an enzyme (5-enolpyruvylshikimate-3-phosphate synthase, EPSPS), which is found in many plants and some bacteria essential to the shikimate pathway to aromatic amino acids.

Thus, the compound (trade named Roundup ™) is a potent herbicide with no toxicity to animals or humans[167] and was first marketed in 1974. The basic patent expired in 1991 and the patent on the preferred isopropylamine salt expired in 2000.

The genomics revolution and the ability to genetically modify seed crops has arrived. Monsanto has also been a leader in providing high-producing hybrid seeds for soy beans, corn and other food crops. The mode of action of Roundup™ was known; and it was discovered that a variant of the key enzyme that was not susceptible to Roundup™ was present in certain bacteria. By genetic techniques, it was a fairly straight forward process to introduce the bacterial gene that is not affected by Roundup™ into food crops. These crops are not affected by Roundup™ but the weeds that grow among them are.

Obviously, this makes the seeds valuable and Monsanto has protected is invention by creating a technology which prevents the "Roundup Ready" plants from reproducing. This "terminator technology" also prevents potential ecological effects of uncontrolled spread of these crops. Overall, this technology has become an amazingly beneficial step forward in agriculture. The only adverse effect of any known consequence is (as completely expected) that natural selection of weeds that can resist glyphosate. This effect is minimized by rotating use of glyphosate annually.

---

[167] It may even lead to antimicrobial discoveries.

Nonetheless a variety of complaints/concerns have been raised. They fall into several categories:

(i) Farmers who have traditionally saved high-yield seeds purchased for one crop for use in next-year's crop are (of course) not happy with the "terminator technology." This is a purely economic issue; the farmers do not want to have to pay premium prices for new seeds every year.

(ii) People who have little understanding of science are concerned because they do not understand what genetic modification means. The term "Franken Food" has been introduced to bring out the idea of the monster "Frankenstein." People want "natural" foods. They do not stop to think that we ingest bacteria, viruses, and a variety of plants and animals (some raw) daily and we do not turn into plants or hybrid animals. DNA from food does not find its way into our genomes or reproductive cells.

(iii) The ever-present threats of cancer, allergy, hormonal disruption etc. can be conjured up (especially in the minds of obsessive-compulsive people). Obsessive-Compulsive behavior can be defined as unreasonable and unsubstantiated fear that results in unnecessary defensive behaviors to the extent that the person's life is limited. There is a spectrum of irrational fear in all of us; and in some situations, strict following of defined procedures may be a desirable trait. But, *excessive* fear of technology, fear of germs and fear of chemicals (chemophobia) seem to play a significant role in the environmental movement.

Primarily because, these people can easily be manipulated and motivated by people with social, political or economic motives.

8.5 Regulation of Food Additives and Drugs

Before 1850, most people (in the US) lived on farms and produced much of their own food. Even city dwellers often had chickens or other small animals associated with their homes. The "tin can" was invented in 1813 in Britain, but it did not become popular until a technique for rapid manufacture was invented in 1846. These were still not consumer items as they were very heavy and hard to open.

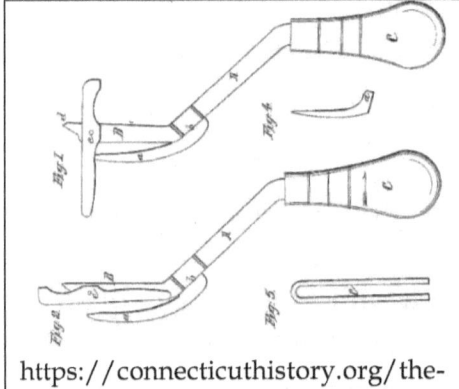

https://connecticuthistory.org/the-first-us-can-opener-today-in-history/

Like many inventions, the American Civil War (1861-65) prompted the development of metal can technology to provide food to troops. This was facilitated by the invention of the can opener in 1858.

The extension of the industrial revolution into the United States during and after the War of 1861-65 involved use of steam engines fired primarily by coal to facilitate fast, long-range transportation and required the concentration of people near factories where large scale manufacturing was accomplished. To feed large cities, food had to be grown in the "country" brought to the city and frequently processed there into cans for even broader distribution (even back to the country).[168]

Thus, an industry sprang up for canning vegetables and slaughtering and processing animals and disturbing the meat in cans. Long-term preservation, of course, became desirable. Refrigeration (<40°F) was not readily available until after 1927. The freezer (<32°F) was not commonly available until after 1945. Both of these advances were facilitated by shifting from toxic ammonia to non-toxic freons as the working gas.

In the 1800s, the main concern of the federal government was in growing enough food to feed the population. As a result, all these activities associated with food were in the Department of Agriculture. The Bureau of Chemistry, in the USDA, became involved in a problem of economic fraud involving the sale of maple syrup products that had been diluted with cheaper forms of sugar and sold in competition with the real product. Harvey Washington Wiley (1844–1930) was a chemist who became an expert in sugar (1887-81) at Purdue University. This experience resulted in an offer to become the chief chemist of the USDA in

---

[168] Nonetheless, in my youth (1950s) my grandfather routinely killed chickens in the back yard for Sunday dinner and neighbors that lived across the road from us (within 5 miles of the NC state Capitol building raised and killed hogs and processed the meat in their backyard in the autumn (when the weather got cold enough for the meat not to spoil).

1882. His initial concerns were sorting out the classification of sugar products and fighting economic adulteration of foods.

The Pure Food Act (1906)

Wiley soon became aware that the issues with commercially processed food went well beyond diluting for economic gain. Things were happening in the food industry that were threatening the safety of the public. Wiley had an ability to appreciate how to get Congress to respond to the needs of the public. In 1902, Wiley established a group of young men who lived in a boarding house and ate foods prepared with substantial doses of various commercial food preservatives (etc.). This group called itself "the Poison Squad" (1902) and Wiley used it to call public attention to the issue of food safety.

Many Sources on the Internet

Wiley's efforts to get public attention were bolstered by Upton Sinclair who published a book entitled *The Jungle* (1905-06) which described unsanitary and harsh working conditions in the Chicago slaughter and meat packing industry. The third pillar of his argument was provided by the women's suffrage movement (1890s, National American Woman Suffrage Association with Elizabeth Cady Stanton). Women saw providing safe food as part of their normal role. And finally, these admonitions did not fall on deft ears. Theodore Roosevelt (a populist politician who was not a favorite of the Republican Party) was run as the Vice-Presidential candidate in 1900 for the second term of William McKinley Jr. (1843–1901). The party leaders felt that this would keep Roosevelt in a largely ceremonial position (and keep him from "rocking the boat") for at least 4 years. They did not count on McKinley being assassinated in 1901. Roosevelt proved to be a president that combined the pro-Americanism of a "rough rider" in the Spanish American War with a nature lover who established National Parks. In 1906, the Pure Food Act passed Congress and was signed by "Teddy" Roosevelt.

The Food Act applied to food in interstate commerce (not local farmer markets) and specifically took aim at "patent medicines," which were typically miscellaneous products (e.g., coal tar, petroleum, herbs) diluted with ethanol. These "medicines" were typically marked as "elixirs" and the term is specifically defined in the Act. Overall, the Act was focused more on *correct branding* of materials sold than on actually setting standards for safety or efficacy. For example, there were *labeling requirements* for alcohol, cocaine, heroin, morphine, and cannabis. For example, "laudanum" (an alcohol extract of opium containing many alkaloids) were freely available. In fact, these materials were taxed in 1914 (Harrison Narcotics Tax Act[169]).

The Federal Insecticide Act (1910)

Farmers are always gamblers. They buy seeds, fertilizer, and insecticides and put these together with their labor in the spring in the hope of being rewarded with a good crop in the fall. We cannot do anything about the weather, but insecticides are chemicals that farmers must buy with no real knowledge that they will or will not work. Thus, in 1910, the Federal Insecticide Act was passed to help protect farmers from buying worthless products that did not kill pest.

And this is where legislations stood until the 1930s.

The Sulfanilamide Episode: Drug Safety

What we think of today as health and environmental regulations did not really begin until 1938.

You will recall that Prontosil was patented and entered the market in about 1933 and in 1936 it gained notoriety by saving

---

[169] "An Act to provide for the registration of, with collectors of internal revenue, and to impose a special tax upon all persons who produce, import, manufacture, compound, deal in, dispense, sell, distribute, or give away opium or coca leaves, their salts, derivatives, or preparations, and for other purposes."

the life of the President's son. This created a general motivation for people to use the drug. Recall that it was not very soluble in anything and generally administered as a pill. And prontosil was never identified with any toxic effects. Nonetheless, when it was discovered that the actual active ingredient is sulfanilamide this compound was long off of patent and anyone could use it.

Meanwhile, mothers were faced daily with their children with a variety of colds and infections. It was very hard to get children (especially young children) to swallow pills. Thus, good marketing suggested that if prontosil or sulfanilamide could be dissolved in a sweet liquid and a flavoring added, children would be much more likely to take the drug and it would sell better at a higher price. It appears that several companies had the same idea.

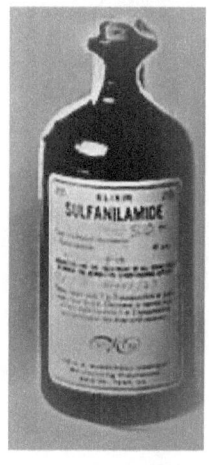

"Elixir Sulfanilamide" formulation[170]: 72% diethylene glycol; 10% sulfanilamide; 16% water; 2% flavoring/coloring syrup (raspberry, caramel and saccharin)

---

[170] A G N. The Elixir Sulfanilamide-Massengill. *Can Med Assoc J.* 1937 Dec;37(6):590.

The S.E. Massengill Company was one of the companies and their chemist did not recognize the toxicity of ethylene glycol…after all it seems like a close relative of ethanol. Thus, the drug was dissolved and sold as "Elixir Sulfanilamide."

The official USDA[171] report reads (in part) as follows:

> "According to the firm's books, 240 gallons were manufactured. The entire amount has been accounted

> Before the "elixir" was put on the market, it was tested for flavor but not for its effect on human life. The existing Federal Food and Drugs Act does not require that new drugs be tested before they are placed on sale.

> Since the Federal Food and Drugs Act contains no provision against dangerous drugs, seizures had to be based on a charge that the word "elixir" implies and alcoholic solution, whereas this product was a diethylene glycol solution. Had the product been called a "solution," rather than an "elixir," no charges could have been brought.

> …

> Most of the drug was administered on physician's prescription."

The report makes it clear that the active ingredient was not at fault and that sulfa drugs in general have been very effective.

---

[171] ELIXIR SULFANILAMIDE-MASSENGILL: Report of the United States Secretary of Agriculture. Cal West Med. 1938 Jan;48(1):68-70.

The report ended with a clear call for new legislation to require safety testing of new drugs. A bill was introduced in Congress in March 1938 and was signed by the President in June. This was the Federal Food, Drug and Cosmetic Act, which is the foundation of the modern FDA.

The Delaney Clause: 1958-1996

In 1958, the Congress attempted to streamline review of new drugs and food additives by creating a category of "generally recognized as safe" for substances with a long history of food use. This simplified the review of compositions that were formulated from a number of different ingredients. However, new food additives (introduced after 1958) were required to pass new safety standards.

But the part of the new legislation that ultimately created the most issues had to do with what at the time seemed like a good idea. By the late 1950s cancer was becoming a bigger public concern than infectious disease (see below). Thus, Congressman James Delaney of New York introduced a clause that affected several sections of the law: (i) food additives, (ii) animal drug residues, and (iii) color additives. The "Delaney clause" was deceptively simple: Cancer causing agents could not be added to foods (as food additives including pesticide residues).

But, this clause (rarely quoted in its entirety) raises several issues including: How do you know something is a carcinogen and how do you detect it? The actual text included a reference to "appropriate animal tests" but rather than litigate this issue no one (to my knowledge) ever tried to argue that any animal test was inappropriate. The other issue of detectability became a very big issue especially with pesticide residues.

Prior to the 1950s, chemists were usually satisfied with detection limits of e.g., 1% (1/100). In 1952 the analytical technique of gas chromatography was introduced and it soon became feasible through extraction and concentration processes and the use of selective detectors to push detection limits for compounds into the part per million (1/1,000,000) range. Ironically, the Electron Capture Detector (ECD) for gas chromatography (GC) was invented in 1957 by James Lovelock.[172] With this tool, samples of the environment (soli, water, air, biological tissues and fluids) can be analyzed to even the part per billion (1/1,000,000,000) level without much effort. The Food and Drug Administration was using this technology in refined forms by 1960. The electron capture detector (ECD) is exceptionally sensitive to and selective for halogenated compounds. Perhaps you recall all of those

---

[172] Ironically, James Lovelock is the creator of the Gaia Theory. This is an absurd notion that all living systems and the non-living (inorganic) elements of earth are combined into one coconsciousness…a single being called Gaia (a goddess). It elevates "mother nature" to god-like status and views any impact on the earth as a sin against nature. It is literally a pagan religious system and has a surprising number of followers, who define themselves in terms associated with witches. They make up a small, but aggressive and extreme, element of the environmental movement.

persistent, fat soluble, chlorinated pesticides that were introduced between 1940 and 1960.

But the first major impact of the Delaney Clause did not involve chlorinated insecticides. It actually involved aminotriazole (a carcinogenic herbicide) used to kill weeds in cranberry bogs.

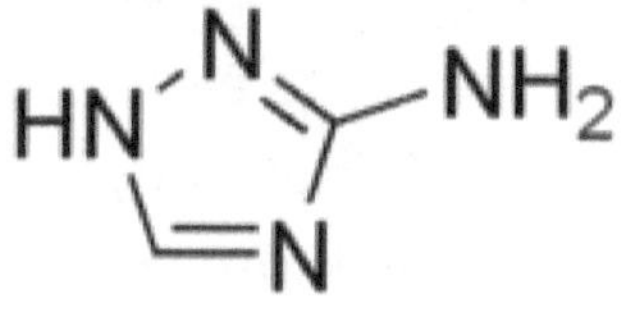

aminotriazole

Source: Wikipedia

A few days before Thanksgiving in 1959, a traditional holiday capped with turkey, dressing and *cranberry sauce*, the Secretary of Health and Human Services (which FDA falls under) announced that cranberries were contaminated in violation of the Delaney clause. And things went downhill from there. The Delaney cause (and the new detection limits and quirks about how carcinogens are defined) created amazing frustration and adverse economic impact because it is almost impossible to use a pesticide on a crop and not have detectable traces of it appear on the product. After years of struggling with the perception of risk, the pesticide residue criteria were removed in 1996 with passage of amendment to Title IV of the Food Quality Protection Act of

1996 (P.L. 104-170, Sec. 404).[173] In the meantime, the market for "organic" foods was created.

Pesticide Regulation

Following WWII, it was clear that DDT was going to be a major business opportunity. And it was clear that there could easily be lots of foreign competition from phosphate esters (parathion) and other products from Europe. Thus, the US pesticide industry was very happy to work with the FDA to develop registration procedures for pesticides in the US that would provide more of a trade barrier to foreign interest than any sort of safety standards in the US.

This may sound like a cynical view, but having worked as a registered lobbyist for a trade association that represented several registered pesticides, I have learned that chemical manufactures do not object to regulations *per se*. In fact, as long as they are provided with (and not excluded from) a "level playing field" they are very comfortable with regulations that everyone must comply with. If everyone is affected, the cost of compliance will ultimately be passed to the consumer (with a profitable markup); *and* the regulations provide a barrier for entry into the market by new competitors.

---

[173]

http://scholarship.law.duke.edu/cgi/viewcontent.cgi?article=1196&context=delpf

Today, merely coming to understand US federal regulations (that affect *every phase* of a start-up business…taxes, payroll, personnel, insurance, environmental, etc.) represents a serious barrier to entrepreneurship.[174] What companies want in regulation is an *absence of uncertainty*, the regulations need to be *clear and unambiguous*. And they need to be *uniformly and consistently enforced*. In this context, the US pesticide manufacturers worked with the FDA to create registration standards in 1947 in the Federal Insecticide, Fungicide and Rodenticide Act (FIFRA).

This act was changed dramatically in 1972 when it was moved to the newly formed Environmental Protection Agency. EPA took over registration and tolerance setting, but FDA inspects and enforces these standards. FIFRA emphasis shifted to environmental and health protection. Pesticides that had been registered were subject to a reregistration process that included the possibility of a *rebuttable presumption against registration* (RPAR). This process went very slowly and there was much litigation. In 1988, more legislation was passed to expedite the process. It was modified again with Food Quality Protection Act (FQPA) of 1996.

---

[174] One of the reasons that immigrants seem to start more businesses than native Americans is the fact that immigrants are frequently ignorant of and find ways to avoid many start-up problems. For example, if you hire only your spouse and children, you can probably circumvent a lot of labor and tax laws. Of course, there are also federal programs (small business quotas for federal spending) that assist the star-up of businesses nominally owned by women and minorities.

Kefauver-Harris Amendment: **Drug Efficacy 1962**

In the early 1960s, the drug thalidomide dominated international news. It had been introduced by A West German pharmaceutical company as a sedative and remedy for morning sickness in pregnant women. At the time, the idea that drugs crossed the placental barrier was not considered of much concern. Regardless, the drug was quickly approved in West Germany for "over the counter" sales (i.e., no prescription necessary). It was also licensed in over forty countries world-wide; but absent appropriate test data, it was not approved in the US. Within two years it was discovered that thalidomide had profound teratogenic effects that involved essentially blocking certain phases of development of the fetus.

Although marketing of the drug was prevented in the US by the existing requirements for safety (since 1938), one manufacturer had declined to proceed with the registration process because they could not confirm the sedative effects that were being claimed in Germany. Thus, passage of legislation (the Kefauver-Harris Amendment) requiring *efficacy testing* as well as *safety testing* was easily achieved. In particular, pregnant women were recognized as a group that required special consideration in safety evaluations.

8.6 The Environmental Movement

Cancer and Cancer Risk Assessment

Between 1905 and 1955, the most dreaded microbial diseases that ended human life prematurely and often with great suffering were largely eliminated. The last major milestone was probably the development of a vaccine for polio. For example, in 1952 approximately 58,000 cases of polio were reported in the USA (mostly children) with 3,000 deaths and 21,000 permanently paralyzed. In 1954, the Salk vaccine against polio became available. Albert Sabin's vaccine came available in 1957[175] and soon polio was an almost unheard-of disease.

As a result, the time was ripe to turn public concerns towards **cancer**. Recall that the Delaney Clause was added to the FD&CA in 1958.

Herman J. Muller and Mutations

After Charles Darwin's (1809-1882) book (1859) convinced most people that there is natural variation associated with inheritance and that this was the basis for evolution, the focus turned to discovering the biochemical elements that accounted for inheritance and variation. Darwin invented the concept of *pangenesis* in which *gemmules* from various parts of the body

---

[175] I got mine on a sugar cube in 1960 in junior high.

were assembled in the gonads (1868, *The Variation of Animals and Plants under Domestication*).

Meanwhile Gregor Mendel (1822-1884) and later Hugo de Vries (1848-1935) developed experimental data on the pattern of inheritance of traits that were useful in understanding the physical system that facilitated inheritance. De Vries published his views in *Intracellular Pangenesis* (1899) and *The Mutation Theory* (1900–1903).

In his view, pangenes were physical objects and this name was shortened to "genes" by Wilhelm Johannsen. Changes in the genes were called "mutations."

Microscopy had improved and the cell cycle was well known by 1900. Walter Sutton (1877-1916) published *The Chromosomes in Heredity* (1903) and Theodor Boveri (1862–1915) reasoned that cancer was an example of abnormal chromosomes and inheritance starting from a single cell. (Before this point, there was no theory of what defined cancer, much less what caused cancer.)

With this basis, Thomas Hunt Morgan (1866–1945)[176] led a research group studying variations in fruit flies (*Drosophila melanogaster*) beginning in 1908. By 1911, he realized that in fact some phenotypic traits were linked to the chromosomes that determined sex; and he generalized this conclusion to the idea

---

[176] E. B. Wilson was a colleague and mentor of Sutton, Boveri and Morgan.

that various traits were associated with genes that were dispersed on different human chromosomes. Later, it became apparent that some traits were physically close to one another on specific chromosomes and by observing when traits tended to be inherited together, a crude map of genes on chromosomes was possible (Alfred Sturtevant, 1913).[177]

Jumping ahead to the 1950s, "genes" were associated with specific codes for making proteins. As a result, individual *traits* have been associated with individual *proteins*. In a few very simple cases this is true. Because some genes make a protein that is critical for a trait. For example, specific proteins are required to make specific pigments and if those proteins are disabled or modified, the pigment either is absent or changed, respectively.[178] But this simplistic system does not account for the major fundamental differences in development and anatomy observed among (e.g., mammals). Mammals include whales, bats, dogs, cows, chimps, and humans *all of which have very similar protein-coding genes*. Obviously, development, involves establishment of a body pattern and subsequent differentiation of

---

[177] It is true that some traits can be greatly modified by a single mutation in a single gene. These are generally, *loss of function* mutations and lost abilities. It does not follow that single mutations of single genes can create entirely new traits (gain of function). Gain of function (especially in development) generally requires changes in how systems of genes are regulated.

[178] Morgan, Sturtevant, Bridges and Muller, *The Mechanism of Mendelian Heredity* (1915).

cells by control of the expression of protein-coding genes. I have called this the "master developmental program"[179] and sooner or later someone will figure out how it works; but it is beyond our current needs to know.

Among Morgan's students was Herman J. Muller. Muller was a theorist and embraced socialist political ideals with sympathy for Marxism as it developed in Russia in 1917. This may or may not be relevant to some of the things that followed. Muller decided in about 1927 to investigate the possibilities of producing artificial mutations in fruit flies. He began with the possibility of using heat, which showed no results; and then toxic chemicals, again with no effect short of lethality to fruit fly larvae. But then he applied x-rays (one of many forms of radiation, as we have seen) and he began to observe mutations. These mutations were typically massive malformations of the chromosomes, which could be observed and documented under the microscope as described in Muller's Nobel Lecture (1946): [180]

> *"In addition to the individual gene changes, radiation also produced rearrangements of parts of chromosomes. As our later work (including that with co-workers, especially Raychaudhuri and Pontecorvo) has shown, these latter were caused in the first place by breakages of the chromosomes, followed afterwards by attachments occurring between the adhesive broken ends, that*

---

[179] Parris GE. Developmental diseases and the hypothetical Master Development Program. *Med Hypotheses*. 2010 Mar;74(3):564-73.

[180]

https://www.nobelprize.org/prizes/medicine/1946/muller/lecture/

*joined them in a different order than before. The two or more
breaks involved in such a rearrangement may be far apart,
caused by independent hits, and thus result in what we call a
gross structural change. Such changes are of various kinds,
depending upon just where the breaks are and just which broken
ends become attached to which. But, though the effects of the
individual "hits" are rather narrowly localized, it is not
uncommon for two breaks to be produced at nearby points by
what amounts to one local change (or at any rate one localized
group of changes) whose influence becomes somewhat spread
out. By the rejoining, in a new order, of broken ends resulting
from two such nearby breaks, a minute change of sequence of the
genes is brought about. More usually, the small piece between
the two breaks becomes lost (a "deficiency"), but sometimes it
becomes inverted, or even becomes transferred into a totally
different position, made available by a separate hit."*

It was found (by Hanson and Oliver)[181] that the incidence of such
mutations was proportional to the *total dose* of x-rays (total dose
= intensity x time):

---

[181] Oliver CP. The effect of varying the duration of x-ray treatment
upon the frequency of mutation. *Science.* 1930 Jan 10; 71(1828):44-6.

Frank Blair Hanson, Florence Heys, and Elizabeth Stanton, The Effects
of Increasing X-Ray Voltages on the Production of Lethal Mutations in
Drosophila melanogaster, *The American Naturalist* 65, no. 697 (Mar. -
Apr., 1931): 134-143.

"Both earlier and later work by collaborators (Oliver, Hanson, etc.) showed definitely that the frequency of the gene mutations is directly and simply proportional to the dose of irradiation applied, and this despite the wave-length used, whether X- or gamma- or even beta-rays, and despite the timing of the irradiation. These facts have since been established with great exactitude and detail, more especially by Timoféeff and his co-workers. In our more recent work with Raychaudhuri (1939, 1940) these principles have been extended to total doses as low as 400 r, and rates as low as 0.01 r per minute, with gamma rays. They leave, we believe, no escape from the conclusion that there is no threshold dose, and that the individual mutations result from individual "hits", producing genetic effects in their immediate neighborhood. Whether these so-called "hits" are the individual ionizations, or may even be the activations that occur at lower energy levels, or whether, at the other end of the scale, they require the clustering of ionizations that occurs at the termini of electron tracks and of their side branches (as Lea and Fano point out might be the case), is as yet undecided."

Muller's comments that (i) the incidence of mutations was independent of wavelength, and (ii) the reason for the multiple breaks in the chromosomes observed from individual hits is easy explained:

X-rays have a wavelength of about 1 nm (or shorter). Recall $E_{photon} = hc(1/\lambda)$. If you plug in the constants and carry through the arithmetic, the energy of an x-ray photon is typically greater than 120,000 kJ/mole.

Comparing the x-ray energy to the typical bond energy of C-C, C-N, C-O, or P-O bonds (which are generally less than 500 kJ/mole) it is immediately obvious that a "hit" from a single x-ray photon would potentially create *several hundred broken bonds* **all at the same time and all at the same place.** These multiple concurrent double-strand breaks allow massive (irreversible) reorganization of the chromosome (as Muller observed). Small changes in the x-ray wavelength would have little effect in the outcome.[182]

Actually, absorption of an x-ray in a single event is unlikely when dealing with light atoms. X-rays tend to dissipate energy by colliding with electrons (of water molecules) producing a spray of beta particles (electrons) and various free radicals:

$$\text{Photon} + H_2O \rightarrow e\text{-} + HO. + HOO. + H.$$

Each one of these particles can be chemically damaging to DNA. In other words, a single "hit" by a x-ray (or other high energy radiation) produces numerous (potentially hundreds) of reactive chemical species (at the same time and same place) and these agents cause multiple DNA strand break and typically complex mutations.

---

[182] If you fall 1000 feet or 2000 feet the results are effectively the same when you hit the ground.

The point here is that a single "hit" from an x-ray is not a single "hit" from a chemical agent. While one "hit" from an x-ray may cause a complex mutation, that is not true for a chemical species.

DNA Repair

At the time of Muller's work and as late as the early 1960s, it was assumed that once damaged, DNA was not repaired. It is now clear that (i) minor damage to DNA is very common[183] resulting apparently from mechanical stress during the many complex movements of these long, thin molecules, and (ii) most mechanically induced breaks or copying errors are readily identified and repaired before or during the DNA is replicates, or (iii) cells that cannot *quickly* eliminate "discontinuities" in DNA typically commit suicide (apoptosis).

On the other hand, errors in which rearrangements of DNA (facilitated *by multiple concurrent* double-strand breaks, e.g., from high-energy radiation) are healed (eliminating discontinuities) incorrectly do not commit suicide and may replicate. We do not have time here to go into the complicated and fascinating mechanism of DNA repair, but I will point out that most tumors seem to be based on some initial chromosome-shattering event

---

[183] These include single-strand breaks and *isolated* (in time and space) double-strand breaks.

called *chromothripsis*.[184] Radiation causes chromothripsis, but radiation does not seem to be the only cause of most examples of chromothripsis found in tumors. It may occur randomly.

Regulatory Standards and Risk Assessment

In the early 1950s, during a period of nuclear bomb testing by the US and the Soviet Union, officials became very concerned about the doses of radiation that *all people* would receive from radioactive "fallout" from dust thrown up with nuclear bombs.

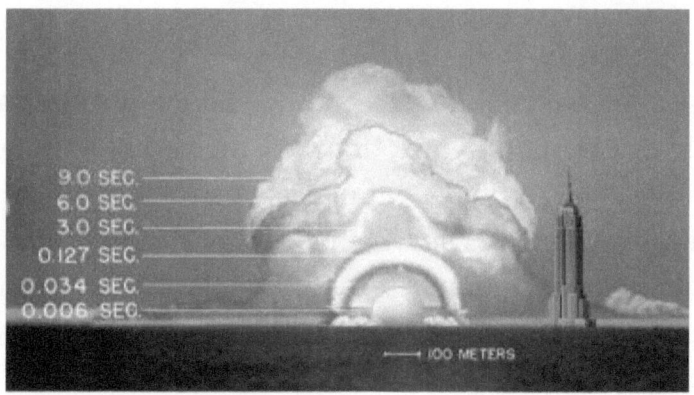

Source: Los Alamos National Lab: 1945 Test

---

[184] Závacká K, Plevová K, Jarošová M, Pospíšilová Š.Chromothripsis - Extensive Chromosomal Rearrangements and Their Significance in Cancer. *Klin Onkol.* 2019 Spring;32(2):101-108.

Bhandari V et al. Molecular landmarks of tumor hypoxia across cancer types. *Nat Genet.* 2019 Feb;51(2):308-318.

Thus, officials concerned with nuclear energy desired to set standards for exposure to radiation. To do this, they first translated all forms of "ionizing" radiation into terms of energy deposited (in a <u>macro</u>scopic mass). The unit was originally the "rad" but it has been eclipsed by the "Gray" which is defined as 1 J/kg (100 rad = 1 Gray). Then based on dose-response experience, they set an acceptable level of exposure. Finally, using the linear, no-threshold (LNT) model advocated by Muller (1946, based on the work of Hanson and Oliver) they proportioned risk to individuals who were exposed to lower levels.

At this point, please note that radiation is not subject to biochemical transformations or pharmacodynamic processes. When a toxic chemical in introduced to the body (orally, via the air or via injection) it typically must traverse a variety of biochemical barriers and survive biochemical degradation and the possibility of excretion before finding its way into a particular cell. In the cell, a chemical agent must traverse the cytoplasm and the nuclear membrane before obtaining access to the DNA. In contrast, radiation or energetic electrons created by deposition of radiative energy goes directly into the nucleus of cells. Even in 1950, this was understood…but ignored when applying Muller's data for radiation to chemicals.

Rachael Carson

We have already mentioned the importance of the introduction of the gas chromatograph with selective detectors as an important factor in motivating and spreading the environmental movement. In order to explain how this played such an immediate role, we need to explain how cancer became a new concern as ordinary diseases became less of a public problem.

About 1960, marine biologist Rachael Carson, who had established a popular following as a nature writer (*The Sea Around Us*, 1951; *The Edge of the Sea*, 1955) developed breast cancer. She had a radical mastectomy in April 1960 and died in 1964. It is likely that her illness contributed to her concerns about uses of pesticides as she wrote a new book *Silent Spring* (1962), which raised alarm regarding the potential adverse effects of pesticides (e.g., bioaccumulation and inadvertent death of song birds). Her book immediately caught the public's attention and initiated a decade of argument regarding the hazard of DDT. DDT was not clearly identified with any human toxicity or carcinogenicity. But it and its degradation products (DDE) was found with the aid of sensitive GC-ECD[185] to be clearly wide-

---

[185] I have already mentioned that in 1952 gas chromatography was invented and in 1957 the electron captor detector (GC-ECD) was available to analyze samples of complex matrices (soil, ground water, air, biological tissues and fluids) for traces of halogenated (e.g.,

spread in the environment and in human and animal tissues. And it correlated with decline in raptor populations (eagles and ospreys), which were found to be producing eggs with thin shells that broke prematurely. The DDT opponents were largely represented by the Environmental Defense Fund (EDF), which was incorporated in 1967. The legal battle over DDT continued until 1972.

In 1970, President Richard Nixon opted to consolidate many of the diverse environmental programs of the federal government into the Environmental Protection Agency (EPA) with the first piece of legislation being the National Environmental Protection Act (NEPA…pronounced Nee-PA). NEPA required the federal government to conduct analytical evaluations and issue environmental impact statements (EIS) to inform the decision makers and the public of reasonable alternatives to major actions and the impacts of those options. It does not tie the federal decision-makers to any particular course of action as long as that course of action is within the scope of the alternatives analyzed.

The National Cancer Act was signed in 1971 and the Nixon administration declared a "war on cancer," which had become one of the leading causes of death in the United States. This act mainly funded much broader research activities. Indeed, it was a war declared without a clear idea of what the enemy was.

---

chlorinated compounds). By 1960 the FDA was equipped with this technology.

The FIFRA regulations were moved from the FDA[186] to the EPA in 1972 with a requirement that the existing pesticide registrations be evaluated under new, much more objective and demanding standards. This process was begun in the EPA-Office of Pesticide Programs (OPP) and proved to be very time consuming and generated much litigation and lobbying. Many existing and new pesticides were declared subject to a *rebuttable presumption against registration* (RPAR) for lack of data or poor quality of data.

Mutations and Cancer[187]

As noted above the association of tumors with abnormal chromosomes went back to the early 1900s. But many unrelated phenomena seemed to *cause* cancer. There were some experiments in the 1940s that suggested that chemical induction of cancer was a two-phase progress: induction and promotion. These observations still did not suggest a mode of causation linked to chemical exposure although it was observed that certain classes of chemicals ('coal tar" "arsenic" and aromatic amines) seemed anecdotally to cause cancer in specific tissues of exposed people.

---

[186] FDA retained the inspection and enforcement duties but EPA set the standards (e.g. pesticide residue tolerances on food).

[187] I have written in some detail about this subject. See: *The Myth of the Linear No-Threshold Dose-Response Relationship for Carcinogens* (2019, paperback, citing 319 papers).

With the revelation of the primary structure of DNA (1953)
theories began to appear in scientific publications linking
mutations to cancer. It is clear from the work of Burdette (1951),
Stigler (1952), Blum (1953), Burdette (1953), Fardon (1953),
Burdette (1954 and 1955) and Nordling (1955) that this is the
period when knowledge of the structure of DNA and Muller's
dose-response hypothesis for mutations were married into the
linear, no-threshold hypothesis of cancer causation.

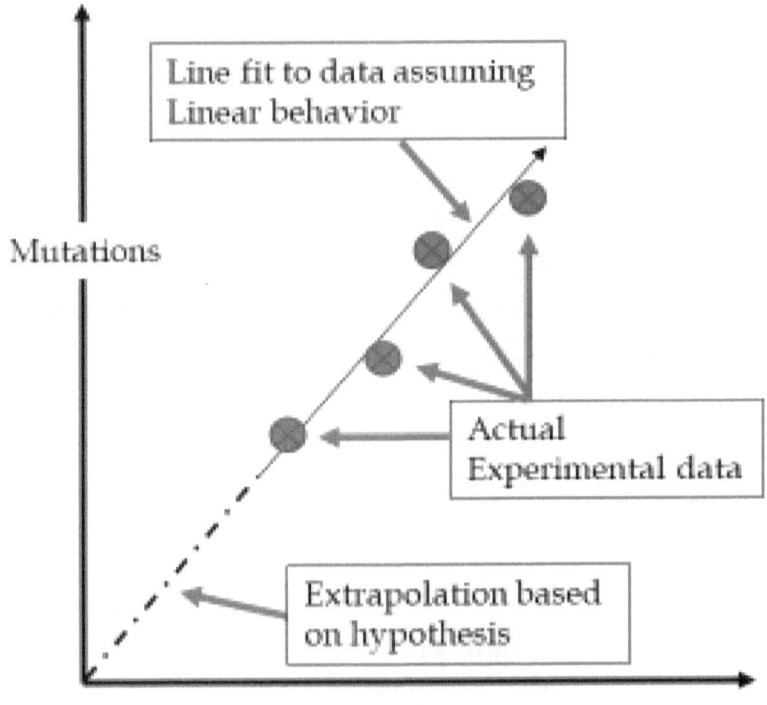

Cancer Risk Assessment and Regulatory Standards

It was not until the US EPA was being tasked to set standards for pesticides and industrial chemical exposure that this theory received a quantitative treatment for chemicals. Remember that EPA was almost exclusively concerned with regulation of chemicals and we have already noted the fundamental differences between mutations caused by x-rays (ionizing radiation) and chemical reactions. Thus, it is very hard to rationalize how Roy Albert[188] makes the following statement[189] in 1994 describing the USEPA logic in the mid-1970s:

> *"Cancer is an expression of genetic damage. Mutations are genetic damage. Cancer is therefore caused by mutation. Mutation is linear with radiation dose in micro-organisms. Therefore, linearity for cancer! The difference between chemical carcinogens and ionizing radiation could be waved aside as they both cause genetic damage."*

On pharmacodynamic grounds alone, "waving aside" the differences between chemical carcinogens and high energy radiation is obviously absurd. Indeed, many mechanisms proposed for carcinogenicity require activation by metabolic processes, which are themselves limited by ordinary dose-

---

[188] Roy E. Albert was well-recognized and considered to be a leading authority on cancer. He was the chairman of the EPA Cancer Assessment Group (CAG) that included a number of prestigious scientists from federal agencies.

[189] Albert RE. Carcinogen risk assessment in the U.S. Environmental Protection Agency. *Crit. Rev. Toxicology*. 1994; 24(1):75-85.

response kinetics and thresholds. The fundamental difference between lesions caused by ionizing radiation and chemical exposure are, of course, ignored in this policy statement as are the effects of DNA repair, apoptosis and the Hayflick limit.

Over the years, various people (including myself)[190] have challenged this policy of the USEPA. Some minor progress has been made, but the application of a simple *linear risk factor* is very attractive among policy analysts (who generally have little background in science or mathematics).

The Risk Standard

In a separate book[191], I have addressed the issue of setting the risk standard for cancer. This is truly a social policy decision, not a scientific decision. And I would prefer that it had been resolved by elected representatives of the people after careful analysis and debate. But it was not. In his 1994 paper, Roy Albert describes a scene in which to bureaucrats met informally in a hallway and one asked the other what is a publicly acceptable risk threshold for a fatal event. The suggestion was that 1 event in 1-million seemed to satisfy the public when deciding whether or not to board an airliner that might crash (in the 1970s). And so, without much thought, a standard of a *total*

---

[190] *The Myth of the Linear No-Threshold Dose-Response for Cancer.* (paperback 2019)

[191] *Acceptable Risk of Cancer: an assay on regulatory policy.* (paperback 2019).

lifetime dose of a carcinogen was viewed acceptable if it only created a 1 in 1-million risk of fatality. I hope you see the error in this logic.

First off it assumes that a total dose of a chemical received over a lifetime is equivalent to a dose received over one hour. There is an obvious difference in "dose rate" in the two scenarios that go back to the pharmacodynamics issues raised above. Part of that confusion arises from the way that typical carcinogenicity bioassays are conducted. In most of these tests, the dose rate is controlled as a constant throughout a fixed experimental period. As a result, the "total dose" is calculable based on "dose rate" times "time."

> **Total Dose (milligram/kg)**
>
> **= Dose Rate (mg/kg/day) x time (days)**

Thus, total dose is proportional to dose rate.

But the real absurdity of this criterion is revealed if we consider other regulatory standards (fully accepted by the public).

Consider traffic accidents.

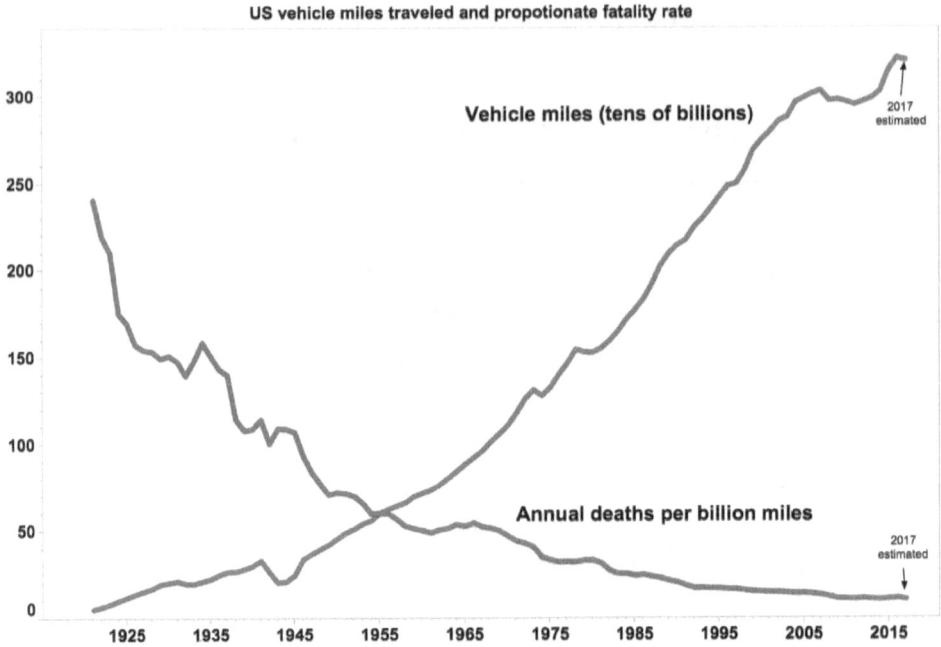

US vehicle miles traveled and propotionate fatality rate

Source: Wikimedia

The death rate in automobile accidents per mile travelled has declined steadily over the years through improved roads and cars (antilock brakes, seat belts, air bags, etc.) but today automobile deaths still stand at about 10 per billion miles. This includes drivers and passengers and does not only consider those who were at fault. You can be perfectly innocently riding in a car and be killed by a drunk driver. Those statistics work out to 1 death per 100-million miles or **1 x 10⁻⁸ death per mile**. By the way, these are *real deaths*, not statistical projections based on some model or hypothesis. Thus, we can predict that if a person drives 100 miles (*in his lifetime!*) his/her risk are

$$100 \text{ miles} \times 10^{-8} \text{ deaths/mile} = 10^{-6} \text{ deaths}$$

In other words, if the Department of Transportation regulated car travel the way that the EPA regulates chemical exposure, no one would be allowed to (legally) travel by car more than 100 miles *in their life time*. Basically, the same government that is accepting a lifetime risk that could easily go to 10^{-3} (0.1%) for normal drivers

$$100,000 \text{ miles} \times 10^{-8} \text{ deaths/mile} = 10^{-3} \text{ deaths}$$

is enforcing a lifetime risk of 10^{-6} (based on an unlikely model) for carcinogens.

Of course, no one wants *unnecessary* exposure even if the risks are minimal. This brings up and important point regarding risk management. If we only consider the risk of exposure to chemicals to be important, we may select life options that will expose us to situations that are actually riskier overall.

I would argue that even if we assume the cancer risk projections by the LNT model are accurate, if we set the risk threshold 1000 time below commonly accepted risk, we will never see any *societal benefit* for the regulations that restrict our use of technology. Of course, if you look only at individual risk someone may be saved by the strict standard…even though they might be killed in an automobile accident later. A risk that we have discounted heavily in recent years is the risk of infectious disease. This has affected funding for research by the government and private investors. More money goes into cancer drug research that goes into antibiotic research. This may turn into a big mistake if we find people dying in the thousands from antibiotic resistant organisms that were easily killed in 1950 by penicillin. It is possible that my generation will be the *unique generations* without fear of plagues; not the *first generation* without fear of plagues.

The Safe Drinking Water Act

The Nixon administration also proposed (1973) an act to ensure that public water utilities provide water that is safe to drink. Legally, the act is only applicable to water utilities of a certain minimum size, which implies that it only applied to systems producing water of adequate volume to meet the daily requirements of a number of families. Moreover, it should be obvious that such utilities would not tap into surface or ground water sources that contain excessive undesirable natural constituents (e.g., salt, hydrogen sulfide, biological activity) that would render the water non-potable on its face. The Safe Drinking Water Act was obviously targeted at manmade pollutants that potentially had adverse effects. These were addressed as primary standards while general characteristics (e.g., color, odor, taste) were left to secondary (nuisance) standards. The general idea was for the federal government (EPA) to develop the standards that the States would enforce through monitoring and publication of results. There was also a public litigation option that was intended to make the act virtually self-enforcing.

When the standards went into effect (June 25, 1977) approximately 40,000 community (e.g., central well providing a group of homes as in a trailer park) and 200,000 public drinking water utilities were covered (not private wells) and they were *required to test the water going to the customers and report noncompliance to the customers.*[192] The regulated constituents included: microbiological contaminants (coli form), ten inorganic chemicals, six organic pesticides, turbidity and radioactivity.

The EPA uses risk assessment (as described above) to develop a public health goal (Maximum Contaminant Level Goal, MCLG) for drinking water. This goal is based on lifetime consumption of nominally 2 liters of water per day by each person (with a assumed live of 70 years)

---

[192] EPA: *"EPA does not regulate private wells nor does it provide recommended criteria or standards for individual wells. ...*Private well owners are responsible for the safety of their water."

There are about 13 million homes on private wells and there could be an issue associated with real estate transactions:
"In real estate transactions, full disclosure typically means that the seller must disclose any property defects and any other important information that could have an effect on a party's decision to enter into the deal." Source: https://www.qualia.com

Most people never test their water; but if they do and find contaminants in excess of drinking water standards, it is rather murky what they should report. Since EPA does not produce standards for private wells, there is no particular basis for "violation" real or implied.

(X) mg/L x 2 L/day x 70 years x 365 days/year =

51 (X) g total dose

and the risk factor for 10^{-6} deaths in the case of carcinogens.

Because of the very conservative risk standard and the (illogical) use of the *extrapolation* of the linear, no-threshold dose-response to very low doses, cancer risks typically determine the regulatory standards. In most cases, acute or chronic toxicity would allow much higher concentrations. It is also clear that many natural water sources contain naturally occurring minerals (elements) that exceed these goals. For example, the concentration of "arsenic" in the Yellow Stone River begins at about 30 microgram/L (i.e., 0.030 mg/L) at the park and is well above 10 micrograms/L far down stream. In such cases and where cost and technology are impractical to require meeting the MCLG (*goal*), the actual *enforceable* Maximum Contaminant Limit (MCL) is selected as near the goal as possible. For example for "arsenic" the MCL is 0.010 mg/L but the MCLG is 0.00 mg/L (based on calculated risk). EPA has provided tables of the current standards: (https://www.epa.gov/sites/production/files/2016-06/documents/npwdr_complete_table.pdf)

An issue that most chemists would immediately raise here is what does the EPA mean by, for example, "arsenic." There are many types of arsenic compounds (organic and inorganic) and it can be shown that they have vastly different acute toxicity and pharmacodynamics. Ultimately, what EPA means by "arsenic" is captured in a codified method for analysis for "arsenic." In other words, "arsenic" for regulatory purposes is defined operationally by a method of analysis.

Overall Risk

There are obvious weaknesses in the US regulatory regime of carcinogens. It is safe to say that in the vast majority of cases, chemicals are over-regulated. That is, the standards are far more restrictive than we would ever expect to see any societal benefit. Indeed, compliance with these standards may actually increase societal risks. For example, remediation of a contaminated site to meet overly strict toxicity guidelines might well require excessive construction activities (construction risk) and movement of soil by truck (transportation risk). As seen by the data above, it would be very easy to incur more transportation risk than would ever be sustained by people living on land slightly contaminated with a carcinogen.

Even among carcinogens there are relevant trade-offs that would be considered in a risk-based environmental strategy. For example, most ground water is contaminated with naturally occurring radon (derived from the radioactive decay of uranium, which has a half-life on the order of a billion years…i.e., it is not going away in the foreseeable future). Thus, suppose we drill wells and extract water contaminated with a man-made volatile hydrocarbon of moderate carcinogenicity. It is relevant that when we extract the volatile hydrocarbon from the water, we are also stripping the volatile radon gas from the water. So far, so good (ignoring the construction and transportation risk associate with implementing the remedial measure and the risk associated with mining coal and generating and transmitting electricity to run the pumps etc.). But now suppose the cleaned water is allowed to infiltrate back into the "clean soil." It will soon be recharged with radon from the naturally occurring uranium and will like have about the same risk as the water before remediation of the manmade contaminants.

As an exercise, calculate the amount of uranium excavated when a typical basement (10 meters x 20 m x 2 m = 400 m³) is excavated. The density of dry soil is easily 1200 kg/m³ and the global average of uranium in top soil is about 1 mg/kg. Thus,

$$400 \text{ m}^3 \times 1200 \text{ kg/m}^3 \times 1 \text{ mg/kg} =$$

$$(4.8 \times 10^5 \text{ kg}) (1 \times 10^{-3} \text{ g U/kg}) = 4.8 \times 10^2 \text{ g of uranium} =$$

$$0.48 \text{ kg of uranium} \approx 1 \text{ pound of uranium}$$

Most readers never have any idea that measurable quantities of natural uranium surround their home.

Thus, this more or less sets a minimum risk that we cannot significantly go below... No matter how hard we try, how much we spend or how much other types of risk we sustaining during our efforts!

Endocrine Disrupters

Realizing that the basis of regulation associated with cancer risk is "shaky" and that medical science continues to improve the prognosis of people afflicted with cancer, it is my opinion that "the professional environmental lobby" has looked for other crusades to follow. In 1996, a book entitled *Our Stolen Future: Are We Threatening Our Fertility, Intelligence and Survival? A Scientific Detective Story* was published (1996) by Theo Colborn, Dianne Dumanoski and John Peterson Myers. The book was complete with a preface by Vice President Albert Gore, Jr. (1948-…) who is also a noted advocate of the theory of manmade global warming. The thesis of the book is that certain industrial chemicals disrupt the endocrine system in unpredictable ways (such that dose-response analysis cannot be applied) and that these effects would likely cause abnormal development especially of sexual systems (e.g., the feminization of males). Those people (in science) with similar (anti-industrial) political views quickly flocked to this hypothesis and a number of papers have been published on various aspects of the topic. And, sadly, a significant amount of *invalid and even falsified* data entered into the literature.

Bisphenol A (BPA), which has structural similarity to estrogen and has some estrogenic effects at high doses, has probably been the most extensively studied "endocrine disruptor" because of its large production and use in epoxy resins and polycarbonates.

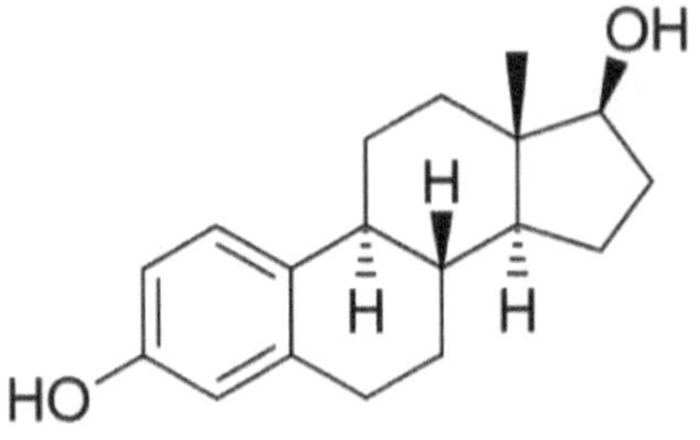

Estrogen
Author: NEUROtiker; Source:
Wilimedia Commons

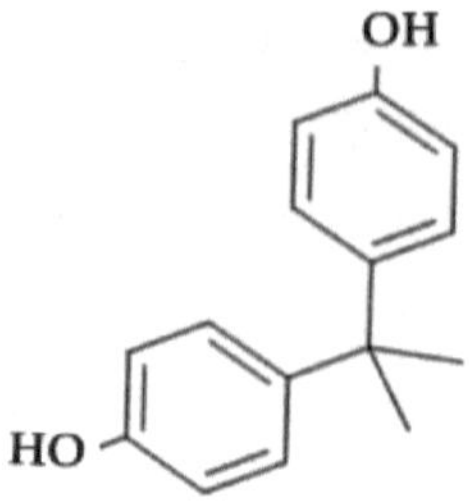

Bisphenol A

Yet in 2009 the following was said about these studies:

> "*In the specific case of bisphenol A (BPA), it is concluded that despite the extensive endocrine disruptor (ED) database available for this chemical, it is still not possible to locate a single study that passes the most rudimentary scientific requirements–that the observations are capable of independent confirmation.*"[193]

Several authors, argued that although in laboratory studies xeno-estrogenic agents (e.g., bisphenol A, DDT) are many orders of magnitude less potent than the actual estrogens and thus play

---

[193] J. Ashby. 2009. Endocrine disruption occurring at doses lower than those predicted by classical chemical toxicity evaluations: The case bisphenol A. *Pure and Applied Chemistry, The Scientific Journal of IUPAC.* Published Online: 2009-01-01 | DOI: https://doi.org/10.1351/pac200375112167.

no real role in health of humans or animals exposed to much lower levels (e.g., as observed in the environment). But the proponents of the theory have argued that effects at "low dose" are actually higher than effects at higher dose (i.e., a non-linear dose response). Such an effect is certainly possible[194], but it is hard to imagine such an effect at the dose levels considered for endocrine disrupters.

If endocrine disrupters are active at low levels, one would expect that gene activation during fetal development would be one key place to look. However, no such effects have been observed for BPA.[195] It is perhaps relevant that I was the principal author of the Test Rule Support Document for the USEPA in 1984 and we reported in that document the meager estrogenic effects of high-level BPA exposure. The EPA never pursued this topic until the book was published in 1996 with the Vice President's preface.[196]

---

[194] Parris GE. 2015. A Hypothesis Concerning the Biphasic Dose-response of Tumors to Angiostatin and Endostatin. *Dose Response*. 2015 May 20;13(2). pii: dose-response.14-020.Parris. doi: 10.2203/dose-response.14-020.Parris.

[195] Aiba T et al. Does the prenatal bisphenol A exposure alter DNA methylation levels in the mouse hippocampus?: An analysis using a high-sensitivity methylome technique. *Genes Environ*. 2018 Jun 4;40:12. doi: 10.1186/s41021-018-0099-y. eCollection 2018.

[196] I was the lead author in 1983-1985. For the Technical Support Documents on Organotins, Bisphenol A and Naphthanate Metal Salts for the EPA-OTS Test Rules Development Branch.

8.7 Case-by-Case Responses to Environmental Issues

Prior to 1976, regulations of chemicals with respect to health effects had been limited to food additives, water contaminants, and releases of chemicals to air or water that produced specific adverse effects.

Leaded Gasoline

Introduction of the internal combustion engine required a compatible fuel. In Part II, we have discussed the broad shifts in petroleum refining first from kerosene (for light) to gasoline (for transportation fuel) and the demand for higher-and-higher octane fuels to achieve efficiency (especially better to power-to-weight ratio for airplanes) in high compression engines. The quick fix (driven especially by war-time expedience 1939-1945) was to add tetraethyl lead (TEL) to gasoline blends starting in the 1920s. But by the 1950s, it was becoming evident that leaded gasoline in passenger cars was causing both a particulate lead contamination in the air of urban areas and a clear build-up of lead in the soil along roadways.

In the 1950s, petroleum companies (Universal Oil Products) were introducing platinum catalysts (invented in 1949) to reform low molecular weight aliphatic hydrocarbons (naphtha) into aromatic hydrocarbon. The aromatics (e.g., benzene, toluene and xylene) had inherently higher octane, but a tendency to produce more soot. In 1973, EPA required a phase out of lead in gasoline and with the mandate for catalytic converters (1975) to reduce smog (caused by unburned hydrocarbons in exhaust) TEL could not be used in these cars. Thus, by 1986 TEL was rarely uses in American cars and it essentially ended in the US by 1996.

Chemical Reactions in the Atmosphere

For our purposes, two zones of the atmosphere are important. The bottom 10-15 km are the troposphere which is characterized by an atmosphere of well-mixed composition 78% nitrogen and 21% oxygen that varies from 760 mmHg pressure at sea level to 200 mmHg at 10 km, and 35 mmHg at 20 km. That implies an oxygen (O_2) pressure of about 7 mmHg at 20 km. Meanwhile, the concentration of water drops rapidly as the temperature falls (see figure). At -10°C (e.g., near the earth's surface) the vapor pressure of water is about 1.5 mmHg, but at -60°C (e.g., at 10 to 20 km altitude) the vapor pressure of ice is about 0.008 mmHg. Thus, the mole ratio of O_2 to H_2O in the stratosphere is on the order of 10,000 to 1. In comparison in the lower troposphere the ratio is only 10 to 1 to 100 to 1. Irradiation of oxygen by short wavelength solar radiation (less than 200 nm) excites diatomic oxygen O_2*and facilitates reactions. In the troposphere, formation of hydroxyl and hydroperoxyl free radicals is predominant; but in the stratosphere reactions with diatomic oxygen are far more likely leading to formation of ozone (O_3). Ozone then "shades" the troposphere by absorbing even longer wavelengths of solar radiation.

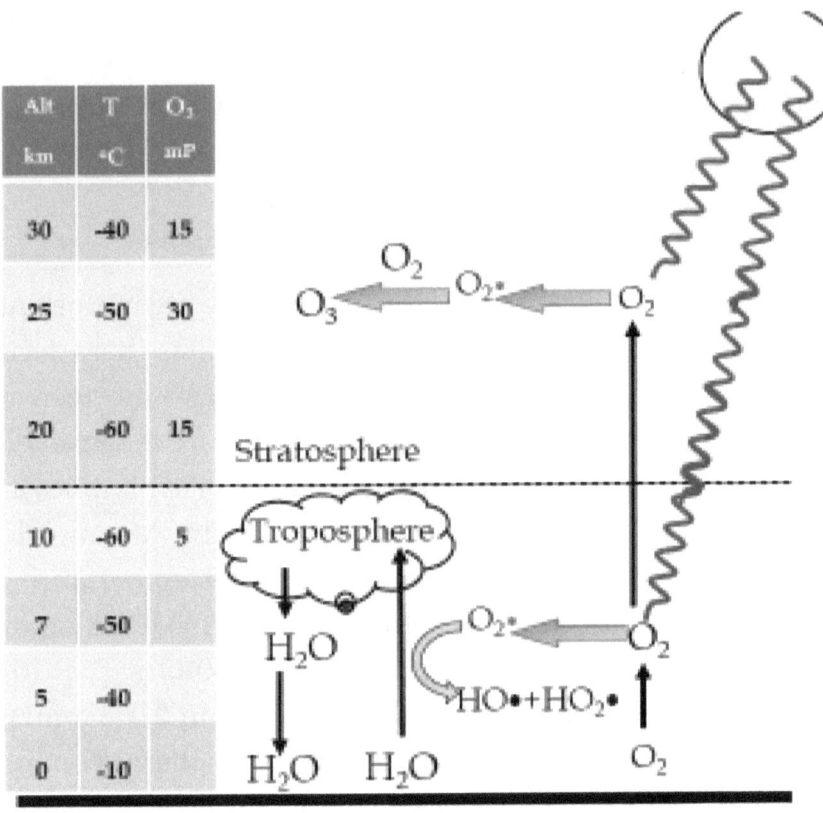

Alt	T	O_3
km	°C	mP
30	-40	15
25	-50	30
20	-60	15
10	-60	5
7	-50	
5	-40	
0	-10	

Chlorofluorocarbons and Stratospheric Ozone

Interestingly, when green plants released oxygen into the atmosphere, they facilitated expansion of habitat for animals onto the land (specifically the open prairie). Solar radiation includes a significant amount of ultraviolet radiation that damages DNA and would have only permitted organisms covered with feathers, hair, scales, or shells to survive outside the shade of trees. Diatomic oxygen (O_2) became a significant component of the atmosphere about 600 million years ago and it quickly dispersed into the stratosphere (12-50 km above the surface) where high energy radiation from the sun facilitates an equilibrium with a second allotrope (ozone, O_3). Ozone absorbs solar radiation at the same wavelengths as DNA; and thus, protects DNA in the skin of animals from excessive mutations and cancer.

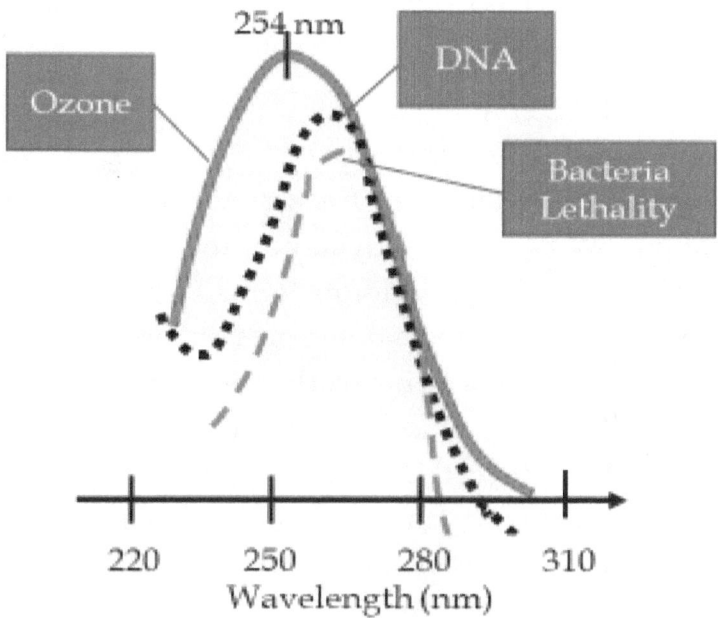

Refrigeration was understood in principle fairly early in the 1800s, but technology did not develop until the 1900s for commercial application of evaporative cooling in a cyclic process to be a practical application. To make this work, a molecular substance that can be liquified under pressure at ambient temperature and allowed to evaporate at a temperature below $0°C$ is required. Ammonia (NH_3) has these properties, but it is lethally toxic and cannot reasonably be used in private homes and public buildings where a leak or puncture could easily cause death. Other compounds (including propane, chloromethane and sulfur dioxide) were also used, but they also were either toxic, flammable or corrosive. Thus, the first refrigerators (invented in 1913)[197] were limited a small market (e.g., commercial ice making).

The Guardian Refrigerator company was acquired by General Motors company in 1918 and renamed "Frigidaire." It was clear that the market for refrigerators, freezers and air conditioning (even heat pumps) would be very large if the technology could be made safe. Thus, General Motors formed a joint venture with Dupont company to develop a nontoxic, nonflammable, noncorrosive working gas. This work proceeded through the 1920s. In 1930, dichlorodifluoromethane and a series of other compounds collectively called "freons" were invented and produced. By 1940 luxury cars were equipped with air conditioners and by the 1960s a majority of cars were factory-equipped with air conditioning.

In an example of unforeseen consequences, gaseous compounds that are not reactive with normal chemical reagents must collect

---

[197] Ironically that was the same year that the ozone layer was discovered by Charles Fabry and Henri Buisson.

in the atmosphere. In 1985, it was realized that chlorofluorocarbons were not degraded in the troposphere and are transferred to the stratosphere. In the stratosphere, they are degraded by high-energy UV to chlorine free radicals, which facilitate the decomposition of ozone to oxygen. This phenomenon was first observed at the poles, but was spreading into the temperate latitudes. Recognizing that this was a serious issue, the manufacturers of chlorofluorocarbons worked with international government agencies (Montreal Protocol) to freeze production in 1986 and reduce production by 50% by 1999. Most production and uses of chlorofluorocarbons were phased out between 2000 and 2010. They were replaced for the most part before 2000 with hydrofluorocarbons (e.g., tetrafluoroethane, CH_2F_4). However, these compounds are under regulatory pressure because they are potent greenhouse gases (i.e., they absorb strongly in the infrared region of the spectrum). They may be replaced with compounds such as ($CF_3CF=CH_2$), which is more susceptible to degradation in the troposphere by attack by hydroxyl radicals ($HO\cdot$).

Why Chemical Companies might want their Product Banned

The case of chlorofluorocarbons reminds me of the case of chromated copper arsenate (biocide) that I was directly involved with and I suspect similar forces were at work in both cases. Consider what happens when an easily manufactured chemical product goes out of patent. Suddenly, anyone can make it and brand it as they like. The product becomes generic and competition drives the profits down to virtually nothing. Competitors typically all belong to "trade associations." And I have noted in several cases where well-meaning but uninformed members of trade associations frequently blurt out something like *"we should all just add a fee..."* or *"we should set a minimum price."* Immediately, the management of the trade association, generally pounds a gavel and makes an announcement (or repeats the first announcement made at the opening of the secession) that *"no discussions of price or cost can be considered here. That would be a violation of anti-price-fixing laws."* Companies that make agreements to set prices are possibly going to be sued or criminally charged.

So, what can an industry do when they are making a great product at a low price and perhaps even losing money and going out of business for their efforts? The most desirable approach is to find a product with lower cost and better properties, which can be patented and branded and sold at a premium. But with something like the original chlorofluorocarbons, the industry was truly in something like an optimum condition of cost/market penetration/benefit. The same situation existed with chromated copper arsenate. *They cannot fix prices, but if* ***government*** *bans the product, it makes it feasible to introduce higher cost products.* Thus, when the EPA or others find a fault with a generic product, it provides an opportunity for industry to accept the banning or other restrictions applied to the use of the generic product so that they can introduce new (not necessarily improved or safer and generally more expensive) products into the market that have the advantage of being *patentable and brandable and thus profitable.*

Competition is ruthless. Capitalism brings the best products at the lowest prices to the market place. Clearly, there are cases where products are found to have adverse toxicity of environmental impacts. But the consumer needs to know that just because a regulatory agency and an industry agree to a course of action, it may not mean that they have compromised for the good of the public. It I paramount that government regulations not be driven by irrational political or religious motivations. Because the result could be products that are more expensive, more dangerous and more expensive than what was displaced.

There are also cases where the government sues a company and the company deliberately does not try hard to win the case. Why? Because the case can be used as legal precedent in future cases against competitors or customers. I'm thinking of a law suite by EPA against a waste disposal company regarding the identification of hazardous waste. By losing the case, the company ensured that much more business was generated by similar enforcement against its customers.

Smog

As discussed above, short wavelength sunlight (UV) activates oxygen in the troposphere:

$$O_2 + h\nu \rightarrow O_2(\text{active})$$

In the troposphere, active oxygen reacts with water to make peroxides and hydroxyl radical:

$$O_2(\text{active}) + H_2O \rightarrow HOO\cdot + HO\cdot$$

Automobile exhaust is not just pure CO_2 and nitrogen. The gases leaving the combustion chamber include nitric oxide (NO) from the oxidation of atmospheric nitrogen, and generally contains some CO and incompletely combusted hydrocarbon in addition to CO_2, nitrogen and even some oxygen. Carbon monoxide, nitrogen oxide (NO) and hydrocarbons react with hydroxyl radicals and hydroperoxides to produce "smog," which is an irritating orange fog. The orange color comes from NO_2.

The primary cure for this phenomenon has been introduction of catalytic converters on automobiles (gasoline engines). The composition of exhaust gases varies from lean (excess oxygen) to rich (excess hydrocarbon and CO). Under lean conditions, CO and hydrocarbons are oxidized to CO_2 and water. Under rich conditions, the reactions favor decomposition of NO to $N_2 + O_2$. There is a narrow range of oxygen concentrations where the catalyst works optimally to eliminate both NO and hydrocarbon.

Concurrent with this effort, evaporation of hydrocarbons from "breathing" storage tanks and displacement of vapors when filling automobile fuel tanks has been implemented.

For engines (diesels) that use heavier fuels, the issue of sulfur in the fuel and complete combustion of carbon (soot) produced in the engine by low-hydrogen content fuels (e.g., a high percentage of naphthalene) produce other issues. The exhaust system of diesel trucks is beginning to look like sophisticated chemical plants that use special (non-fuel) additives to process the engine's combustion products to minimize air pollution. When low-sulfur fuel is used, the exhaust typically goes initially into a catalyst to oxidize the diesel fuels. This process produces soot which is then removed in a filter (which must periodically be changed). The gas then has an ammonia additive injected and passes to a second catalyst where NH_3 and NO react to form N_2 and H_2O.

In Summary

In summary, the three cases described here ((i) leaded gasoline, (ii) chlorofluorocarbons, and (iii) smog) are examples of situations where beneficial products proved to have adverse effects when used on a large scale. Other situations such as the foaming caused by non-biodegradable surfactants (e.g., branched-chain alkylbenzene sulfonic acids) could have been mentioned. And (as of 2020) there seems to be an emerging issue with water-soluble perfluoro-octanol and related compounds. These situations, in some cases, can be (and should have been) anticipated; and in other cases, the problem is purely a matter of scale, which was never anticipated. Regardless, in the US, laws have been passed that attempt to identify and correct these issues early and that is the next topic.

8.8 Systematic Regulation of the Chemical Industry

We have already discussed food additives, drugs and pesticides. These are obvious areas where regulatory oversight by government is desirable. Now, we will look at attempts by government to anticipate and head-off problems before they become big problems (i.e., expensive and disruptive to correct).

Toxic Substances Control Act

In the 1970s, momentum was clearly moving towards general regulation of the chemical industry. And in 1976 Congress passed and the president signed the Toxic Substances Control Act. Under this act, the entire chemical industry was regulated with a presumption of harm.

It is relevant that the passage of this legislation (1976) [198] followed a pattern of other related legislation in that an example had to be

---

[198] I graduated from Georgia Tech in 1974 with a PhD in organic chemistry and the goal of obtaining an academic research position. Little did I know that I was going to spend the next 30 years engaged in (and a witness to) the environmental movement. I found a post-doctoral position at the National Bureau of Standards (now the National Institute of Standards and Technology, NIST). The group under Frederick Brinckmann was interested in bio-methylation of metals (e.g., mercury, tin, lead). I was (and still am) one of the few people in the world who has actually worked with methyl compounds

presented to stir public interest/alarm. However, in this case there was no immediate health related episode. It turns out that while looking for DDT and other chlorinated pesticides, a frequently larger and complex pattern of peaks was observed in GC-ECD chromatograms of environmental samples. It turned out to be the result of a product called PCB (polychlorinated biphenyl). This material had been "flying under the radar" since the 1920s. The product is made by chlorinating biphenyl until a certain degree of chlorination is achieved (e.g., weight percent chlorine, usually on the order of 50-70%) and the product had the desired viscosity.[199] It has very low solubility in water and generally low toxicity, but it bioconcentrated in fatty tissue of animals and humans. Its use grew with the spread of electricity because it has good electrical insulation properties and it is not

---

of arsenic and antimony. This was an ironic twist of fate that I had worked as a lab assistant for G.G. Long at NC State University as an undergraduate for 4 years. My work at Georgia Tech under Eugen C. Ashby had been with organo-magnesium compounds that were not of environmental interest. Towards the end of my post-doc, NBS could not hire me, but sent me on detail to the USEPA Office of Toxic Substances that numbered about 40 people working on a variety of projects (through contractors) to support the passage of legislation in Congress. It was quite a shock for a person who was only interested in science to find myself surrounded by people who were mainly concerned about politics.

[199] There are 10 compounds ($C_{12}H_xCl_{10-x}$) and each compound can have multiple isomers because of different substitution patterns on the aromatic ring.

flammable. Thus, it seemed ideal for use in electrical transformers and other electrical applications. It was also useful as a heat transfer fluid. Ultimately, enormous quantities were produced and spread wherever electrical transformers were used. And it was persistent in the environment. Thus, by the 1970s it was found in many places. It was used as the "poster child" product to support arguments that there should be a regulatory regime that considered industrial chemicals more or less with the same safety considerations as developed for food additives in 1938. (PCBs is the only product specifically named in TSCA.) But in the case of TSCA, EPA would be able to limit or ban a product before it was ever manufactured (or imported) into the United States or EPA could limit the amount produced and the uses (markets) it could be allowed into.

Thus, TSCA divided the chemical world into *old chemicals* (in commerce before the TSCA, 1976) and *new chemicals* (anything after TSCA). The first task of the office was to establish an inventory of chemicals in commerce. The universe of old chemicals was established by sending out questionnaires to chemical companies and having then identify products, by Chemical Abstract Service (CAS) number, that they manufactured or imported. Data were also gathered regarding amounts and uses. This was all compiled into the initial *TSCA Inventory*. To a chemist, this looked a little strange for the following reason. There are CAS numbers for mixtures of materials that are not fully characterized and which vary within ranges. Thus. a compound such as "benzene" appeared as a specific listing in the TSCA inventory (CAS 100-41-4)[200], but might also be a major component in other products identified by other CAS numbers.

TSCA can be subdivided between *data gathering* and *regulatory action.* When TSCA was passed and implemented, there was *no other* federal legislation that addressed waste management or response to "releases of hazardous substances into the environment." In principle, TSCA could have addressed all contingencies for industrial products "from cradle to grave." There were already some federal laws dealing with consumer products (Consumer Product Safety Commission) and worker health and safety (Occupational Health and Safety Act, OHSA), but these acts and the organizations that implemented them were generally not as potentially restrictive (or potentially punitive) as TSCA. Regardless, there was a section introduced into TSCA that essentially says that if any other federal law can regulate in a specific area, TSCA is to defer to that law/agency for remedy. Obviously, materials used as pesticides fell under FIFRA, materials used in medicine fell under drugs (vaccines and medical devices) FDCA and anything intentionally put into food fell under FDCA (see above). Nonetheless, if a compound was registered as a pesticide also found use in some other industrial application, it would be regulated "cradle to grave" for the other uses under TSCA. Thus, my colleagues at the Office of Toxic

---

[200] The last digit is a "check digit". To calculate the check digit that the other digits and multiply be the corresponding integer (1, 2, 3, …right to left). Thus, for benzene: 1x5 + 0x4 + 0x3 + 4x2 + 1x1 = 5+8+1 = 14…check digit 4
Bet you didn't know that.

Substances (1976-77) speculated that TSCA could give rise to an entirely new and separate agency. They were wrong; Congress was already working on other laws (see below).

In the end, TSCA has a role similar to FIFRA for industrial chemicals. Entry into the TSCA inventory is comparable to becoming a registered (allowed) product. As seen in the case of FIFRA, the *old chemicals* were subjected to continual reexamination. If the EPA had determined that a chemical is allowed for a specific use (but not others), a manufacturer is required to file for a "Significant New Use Review (SNUR)." New chemicals often have the specific uses spelled out in the review process.

The review process under TSCA involves many of the same skills used to review pesticides under FIFRA and the two laws ended up under the same administrative offices (the Office of Toxic Substances and Pesticide Programs) and this has involved farther as new regulatory authority was added (currently: Office of Chemical Safety and Pollution Prevention (OCSPP) which includes: the Office of Pesticide Programs, the Office of Pollution Prevention and Toxics; and Office of Science Coordination and Policy). The "pollution prevention" concept largely is associated with the idea of "green chemistry."

Mechanistically, here is how the old and new chemicals are processed:

Old Chemicals are periodically screened (frequently by contractors) based on production volume and likely human exposure and a ranked list is proposed to an "Interagency Testing Committee" made up of experts from several federal agencies. The selected chemicals are ranked by the ITC and they request (via contractor) dossiers on each chemical to be produced.[201] These documents were typically less than 50 pages and included sections on production and use, environmental fate and impacts, metabolism and various toxic effects (acute, chronic, cancer, teratogenicity, etc.). The main consideration by the ITC was whether or not the chemical had adequate testing data to understand its hazard. Those chemicals that were missing data, we referred to the OTS-Test Rules Development Branch for development of Test Rules. Again, contractors[202] prepared much more detailed support documents (hundreds of pages) which were intended to support the EPA rulemaking requiring testing of the chemicals for various criteria. During this process, the manufacturers were contacted for any unpublished information and EPA generally invited them to come in and informally discuss the direction that EPA was headed. In many cases, upon learning of the "data gap" that EPA had identified, the affected industry would voluntarily initiate new testing or produce data that were relevant to the questions. Of course, much of this had to be managed as Confidential Business Information (CBI). Ultimately, if the EPA

---

[201] This was the first project I worked on as a contractor to the ITC. It was a great opportunity to see the broad scope of chemicals in commerce.

[202] In the 1980s, I supported this process as well.

was not satisfied, a test rule was promulgated (with public comments and responses) via the *Federal Register*.

For *new chemicals*, the company submitted a "Pre-Manufacturing Notice" (PMN), which was subject to both an economic analysis and technical analysis. The economic analysis was intended by EPA to estimate the size of the market (hence the production volume and uses). To do this, EPA (again through contractors[203]) estimated manufacturing costs and looked for markets that might not have been considered by the applicant. [204]

---

[203] Yes, I supported this program as well.

[204] In my experience, this was one of the weakest elements in the EPA program. The people within EPA generally had no understanding of the chemistry (e.g., how the compound was actually made) and thus could not have a clue how to estimate the cost of manufacture for a specific manufacturer. One of the most memorable episodes I was involved in was a PMN that was actually a waste stream (still bottoms) from the manufacture of a specific product. This waste/by product stream would probably have qualified as a hazardous waste under RCRA (see below) which would have been expensive to dispose. But it was being recommended in a generic use in a bulk market. When I pointed out to the reviewers at the EPA that (i) the production of the material (by product) would be limited by production of the main product and (ii) if approved by EPA under TSCA, it would effectively have *negative* manufacturing cost (since *it was already being produced* and incurred cost of disposal) I was met with ridicule. [How can a product have negative manufacturing cost and a limited market?] These conflicting ideas were beyond the ability of the EPA managers (who

As I stated, TSCA is a safety net that catches almost all possibilities. It has some specific exclusions (e.g., tobacco, radioactive materials) and in the end Congress agreed to exempt research activities in academia and industry. (Can you imagine trying to conduct research in an environment where every experiment had to be reviewed by government bureaucrats!) But TSCA was very quickly followed by RCRA and CERCLA (see below), which effectively address waste management and improper waste management.

The idea of eliminating wastes and increasing safety by substitution of safe chemicals for hazardous chemicals was imbedded in the TSCA program from its inception. Indeed, I managed a project for EPA in the early 1980s to consider alternatives for epichlorohydrin, whose main uses are manufacture of glycerin and epoxy resins. We considered alternative routes to epoxy resins and the cost and waste associated with these processes in a report published by the EPA in 1984.

Epichlorohydrin is used as an intermediate (i.e., no end uses). As lead author on this report (1984), I and other chemists and chemical engineers working for me at Dynamac Corporation looked into alternative pathways for the principal use of epichlorohydrin in epoxy resins.

---

were neither chemists nor economists) to understand. There were actually several situations like this participated by the hazardous waste legislation (see below).

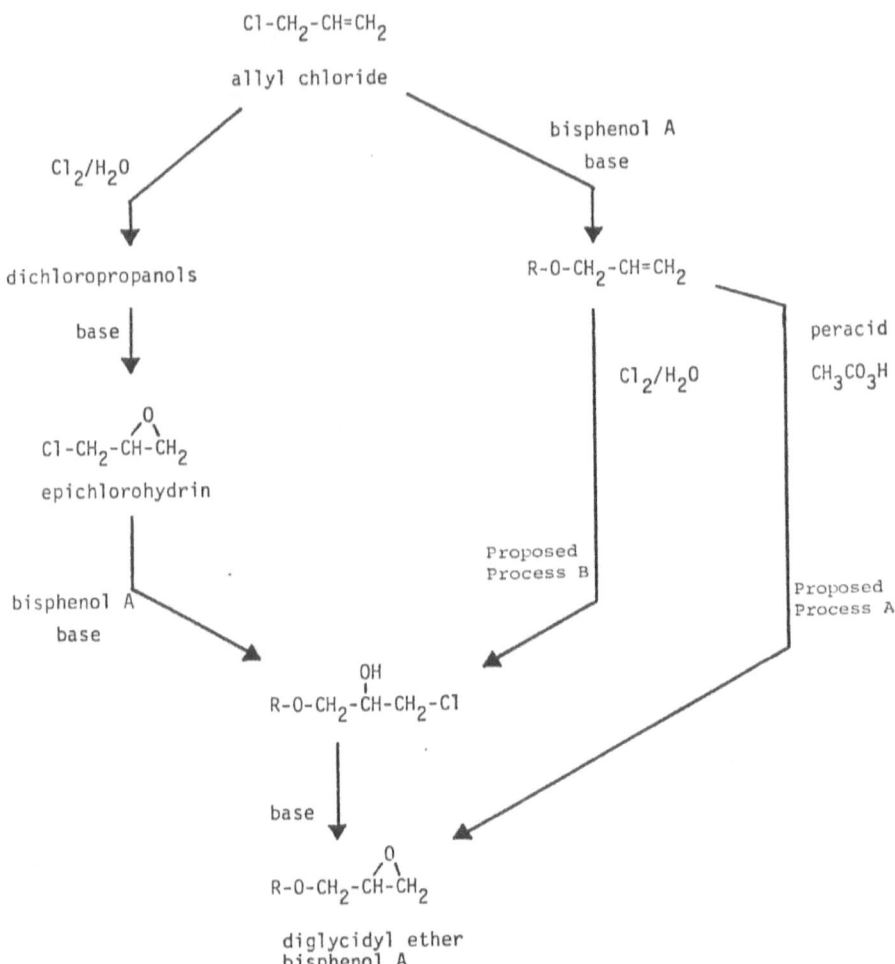

We concluded that there were two alternative pathways to manufacture the same end product (i.e., the diglycidyl ether of bisphenol A) and examined the cost of production and waste generated in each case.

Table 4.2.3A. Raw Materials Needed to Produce One Pound of DGEBPA

	Cost[a] $/pound	Conventional Process pounds[b]	cost($)	Proposed Process A pounds[b]	cost($)	Proposed Process B pounds[b]	cost($)
Allyl chloride (FW 75)	0.61	0.82	0.50	0.61	0.37	1.0	0.61
Chlorine (FW 71)	0.08	0.77	0.06	--	--	0.53	0.04
Caustic (50% solu.) (FW 40)	0.15	0.87	0.13	--	--	0.79	0.12
Bisphenol A (FW 228)	0.67	0.74	0.50	0.93	0.62	0.82	0.55
Peracetic acid (35%) (FW 76)	2.66	--	--	0.61	1.63	--	--
Potassium carbonate (FW 138)	0.34	--	--	2.29	0.78	--	--
Acetone	0.21	--	--	0.15	0.03	--	--
Methylene chloride	0.29	--	--	0.15	0.04	--	--
Methyl Isobutyl Ketone	0.47	0.09	0.04	--	--	0.09	0.04
DGEBPA product:		1.0 pound/$1.23		1.0 pound/$3.47		1.0 pound/$1.36	
Solid wastes produced:[c]		2.29 pound		3.74 pounds[d]		2.23 pounds	

[a]Based on current prices for large volume purchases.
  Data for peracetic acid are from FMC. Other data are from CMR (Hammaker, 1984).
[b]Based on Dynamac chemical and engineering estimates.
[c]Disposal cost are not considered here but could be about $0.05/lb.
[d]About one pound of this is CO_2 vented from the reactor.

Figure and Table from the Report. Cost figures were for 1983-84.

Based on 1983 cost data, we considered the peracid epoxidation of the diallyl ether of bisphenol A to be prohibitively expensive. But, over the last thirty years several new methods of epoxidation have come forward that would greatly improve the economics and perhaps reduce the byproduct waste generation (see below). Nonetheless, to my knowledge, EPA has taken no action on epichlorohydrin. The traditional route via epichlorohydrin may still be the most cost effective overall, because epichlorohydrin can also be used to make glycerin.

Thus, I was somewhat blasé when I heard (a decade later) that EPA had a *new* program for "green chemistry" (developed between 1995 and 1998).[205]

Resources Conservation and Recovery Act (1976)

In the late 1970s and early 1980s, there were a number of news stories that were basically sending the messages (i) that we were running out of space for landfilling waste, (ii) that landfills were polluting ground water, and (iii) that incinerators were polluting the air. We had also gone through a petroleum (gasoline) shortage in the 1970s that heightened everyone's concern about the exhaustion of natural resources. Thus, the idea of conserving and recovery (e.g., reuse, recycling, and waste minimization) of materials was in most people's minds. These ideas were being channeled into legislation by the EPA and Congress. The result was legislation specifically targeted at waste management: The

---

[205] I guess it is all in the marketing and perhaps I should congratulate Paul Anastas the "father of green chemistry." By the way, I worked with Joe Breen (1942-1999) at EPA (1976-78) who is remembered as "the heart and soul of green chemistry." I also shared an office at EPA with Joe Seifter (1904-1982). He took me over to watch at experiment at FDA with arsenocholine by one of his students.

Resources Conservation and Recovery Act (RCRA), which followed closely behind TSCA. The Act was significantly amended in 1980 and 1984.

The Act defines "hazardous waste" as "*a solid waste . . . which because of its quantity, concentration, or physical, chemical, or infectious characteristics may (A) cause, or significantly contribute to an increase in the mortality or an increase in serious irreversible, or incapacitating reversible, illness; or (B) pose a substantial present or potential hazard to human health or the environment when improperly treated, stored, transported, or disposed of, or otherwise managed.*"

This provided an interesting example of the way that laws passed by elected officials are implemented by unelected bureaucrats:

Implementation of RCRA was left to the US Environmental Protection Agency (EPA). The language of the Act instructs EPA to "*…promulgate regulations identifying the characteristics of hazardous waste, and listing particular hazardous wastes (within the meaning of section 6903(5) of this title) …*" (USC 42 § 6921(b)(1)). Although, this language suggests to me that EPA should (a) develop defining characteristics of "hazardous waste" and (b) lists those wastes that meet those characteristics; EPA took an entirely different approach: The USEPA (A) defined characteristics to identify "hazardous waste," and (B) where gaps in the characteristics might exist in EPA's opinion, certain waste streams have been *identified operationally* as "hazardous waste." Whether it intended to do so or not, EPA has created a dual structure for identifying hazardous waste (40 CFR 261.10

characteristically hazardous wastes and 40 CFR 261.11 *listed hazardous* waste). In my view, this approach has no basis in the mandate provided by Congress and greatly overreaches the objectives of the legislation as passed by Congress. Specifically, in EPAs interpretations: **listed hazardous wastes are not necessarily characteristically hazardous**; in fact, they generally are not characteristically hazardous. The first rules appeared in 1980 as the EPA program became effective.

Consistent with the directions from Congress, the first method devised to identify hazardous waste was to set up objective tests that measured chemical, physical and toxicological characteristics that were linked to the ability of the waste to have the adverse impacts that Congress wanted to protect against (40 CFR 261.20 - .24). One of these tests was originally (1980) an Extraction Procedure designed to simulate what would happen if a waste were managed in an ordinary solid waste (municipal) landfill (i.e., as a non-hazardous waste). In this test procedure, the concentration of a few constituents for which federal drinking water standards (i.e., Maximum Contaminant Levels for public drinking water, MCLs) had been developed (under the Safe Drinking Water Act) were measured in the extract. The RCRA standards were obtained by assuming there would be some dilution or attenuation of the leachate before it could become drinking water and the Extraction Procedure for characteristically identifying hazardous waste used multiples of the drinking water standards (e.g., 100 x the MCLs) as the hazardous waste criteria. This entire concept was extended in the mid-1980s with the Toxicity Characteristic rule in which a longer list of constituents of concern was prepared and a more elaborate leaching test (the Toxicity Characteristic Leaching Procedure, 40 CFR 261 Appendix II) was developed. All of this makes logical scientific sense. Although we can debate whether the attenuation factors used to relate the risk-based drinking water standards to the RCRA criteria are appropriate or whether the leaching procedure is too aggressive or not aggressive enough; the logic of the test follows from what happens in the real world and *you can objectively (i.e., experimentally and scientifically) decide if a non-excluded solid waste is hazardous without a waste manifest or an attorney*. **This turns out to be very important**.

Obviously, if a solid waste fails any of the characteristic tests, it is a hazardous waste, except for some solid wastes that are excluded from being hazardous waste for various technical reasons (40 CFR 261.4(b)). In general, objectively measurable characteristics are an attractive way to classify hazardous waste, not only because you know when you enter the waste management system (40 CFR 261.3(b)), but also you know when you exit (40 CFR 261.3(d)(1)) *without filing any reports, submitting petitions or waiting for government decisions or rulemakings*.

Unfortunately, the characteristic approach had two weaknesses: (1) The original Extraction Procedure covered very few constituents and even the newer Toxicity Characteristic rule does not cover the entire universe of possible hazardous constituents, and (2) the characteristic is potentially susceptible to "sham treatment" (typically dilution). The EPA thought it would solve both these problems by providing lists of certain waste streams that the EPA has *typically found to be hazardous through testing similar to the characteristic tests.* In this concept, EPA does all the testing and sustains the burden of proof to demonstrate through the rulemaking procedure (including a regulatory impact analysis to identify the financial impact of the cost of regulation, EO 12291) that a waste is a hazardous waste. This sounds good to industry, and the EPA accomplishes its objective by defining the method of generation (an operational definition of the hazardous waste) and declaring that the waste stream must be tracked through manifest and managed as a hazardous waste without regard to normal variations in the exact constituents or their concentration. EPA could have completed the job in 1980, at this point, by promulgating a rule saying the listed wastes could not be diluted or treated before testing for the characteristics. But, in 1980, this approach apparently seemed unenforceable. Nonetheless, EPA was ultimately (1984-86) forced to resort to such a rule (40 CFR 268.3) and the world has not come to an end.

It is worth bringing up a semantic point that has plagued this field since its inception. Namely, what do we mean when we say "waste stream." Originally, the waste stream may have meant the "point of generation" of the waste, but early on, the term was adapted to mean the waste itself: "substance generated." *Thus, the waste substance was imbued with a non-chemical intrinsic property of being "listed."* Unfortunately, it takes political and legal alchemy to remove the stigma of "listing" from materials; <u>chemists are of no help</u>.

Thus, with good intentions, the **listed hazardous wastes** were born. But, unlike the situation with characteristically hazardous wastes, if the waste is listed (or trapped in the listed category by certain rules discussed below), you cannot get out of the hazardous waste management system without a formal petition (40 CFR 261.20) and lengthy rulemaking (40 CFR 261.3(c)-(d)(2) and 260.22) called "delisting."

Remember the key to the rationale for a system of listed hazardous waste was that the operationally defined waste streams were specifically studied by the EPA. The burden of proof had been *on the Agency* to show through the rulemaking process that the operationally defined waste stream met the criteria set out by Congress. In principle, under this situation, the waste *at the defined point of generation* could be fairly certain to fail the existing characteristic test (or a characteristic test could be devised if it did not already exist); and through the listing process and the requirements for manifesting, these wastes would not be lost from the management system.

EPA proceeded (1980) to establish lists of waste streams from non-specific sources (e.g., still bottoms from recovery of various types of solvents, 40 CFR 261.31), waste streams from specific product-processes (e.g., recrystallization liquor from a specific type of reactor at a specific point in the manufacture of certain products by a certain process, 40 CFR 261.32) and a list of commercial products that become hazardous wastes when disposed (40 CFR 261.33). There are a variety of features to these lists that are worth comment:

♦Remembering that the operationally defined waste streams had to be shown to meet the Congressional definition contained in RCRA, it is interesting that the operational definitions published in the list of waste from non-specific sources themselves often have concentrations in the definition. For example, several of these listed hazardous waste definitions (e.g., waste streams designated F001 - F005) read as follows, "*The following ....[spent solvents]..containing, before use, a total of **ten percent or more (by volume)** of one or more...[solvents]...; and still bottoms from the recovery of these spent solvents and solvent mixtures.*" From these definitions, it is clear that the wastes that were being deemed to be hazardous contained *percent-levels* of the hazardous constituents and that *concentration was critical to the argument* used in the rulemaking for adding them to the list. The preambles to these listing rulemakings typically contain language as follows, "*EPA has determined that these wastes contain toxic constituents **at concentrations** which pose risks which are unacceptable... .*"

♦With regard to the identification of hazardous constituents and development of adequate test methods to monitor them, there is no conceptual reason that an exhaustive test protocol cannot be developed. Indeed, the work done by the EPA in developing the Hazardous Waste Identification Rules (published in 1995 and 1996) essentially establishes "bright line" concentration limits (i.e., "characteristics" by another name) for virtually every constituent of concern in waste streams and media. So, it is technically feasible to establish *objective characteristics* for all *real hazards.*

♦ In practice, the U.S. EPA has its share of zealots such that the relationship between an industrial waste stream being listed as a hazardous waste by the EPA and any reasonable expectation that it might actually be (characteristically) hazardous is sometimes pushed beyond the limit. For example, a memo dated 13 February 1987 from the EPA "Wood Preserving Hazardous Waste Listing" Project Officer [Dr. Cate Jenkins] to the Chief of the EPA Waste Characterization Branch and the Chief of the Listing Section concerning the "Wood Preserving/Dioxin Hazardous Waste Listings" says a lot about the way the EPA operates. The EPA staffer argues that although (a) no hazardous constituents are found in some of the waste streams and (b) in some cases the constituents that have been found have not been shown to be toxic, this should not stop EPA from listing the waste stream as hazardous. The following quote (from Cate Jenkins) contains an incredible admission of an Agency run amuck:

..."*If either of you have problems with any of the* [planned] *listings, you should inform me either verbally or in writing. I will document conversations as I have in the past and formally present any rationales.*"

"Note that I would draw heavily from Agency precedence in this respect. One particular key listing would be the industry non-contested proposed listing of UDMH intermediate overheads, which have never been found to contain the toxicant of concern, UDM[H], and which furthermore would not be expected to contain this toxicant, since the final UDMH synthesis step was not even performed at the point this waste was generated. We were concerned about potential back-flush contamination from up-line processes. Another particularly appropriate precedent setting listing is Hazardous Waste No. F026, Wastes... from the production of materials on equipment previously used for the production of materials on equipment previously used for the manufacturing use...of tetra-, penta-, or hexachlorobenzene...[This is a staffer typo, not agency double-talk]...I challenge you to find any supporting analytical data that [we] were able to use to support this listing."

The bureaucrat (Cate Jenkins) is bragging that the Agency (under her guidance) has listed waste streams without any supporting evidence and argues that this approach is precedent for more unsupported listings!!! This is a victory of environmental extremism over reason, which clearly shows how the US EPA can and has used the listing process to arbitrarily regulate industry.

♦ In an ironic twist of fate, EPA argued in the 1990s not to add certain solvents to the 40 CFR 261.31 list (*Federal Register* 61(158) 42319) as follows:

"…This [decision not to add these spent solvents to 40 CFR 261.31] is not a determination that these chemicals are nontoxic. Many of these solvent wastes are, in fact, already regulated as hazardous waste because they exhibit a hazardous waste characteristic …and/or because they are mixed with other solvents or wastes that are, themselves, listed hazardous waste…"

Apparently, EPA has digressed to the point that they are only listing NONTOXIC materials as hazardous waste! At least EPA seems to agree that the existing characteristic tests are, in principle, adequate protection and that no additional regulation through listing is necessary. [Don't tell the EPA, but many of the spent solvents that they are proposing not to list in this case, are NOT covered by the current toxicity characteristic rule 40 CFR 261.24.[206]

♦ Once a waste is captured by the "listing" any relationship between the hazard of the waste and the requirements for its management are lost. For example, in the 1986 land disposal standards (40 CFR 268.40) some wastes must be treated below the characteristic levels before they are land disposed. Indeed, because of the

---

[206] The real reason that they were not promulgating a listing was that they had applied risk assessment technology developed in the contemporaneous Hazardous Waste Identification Rule process and discovered that the risk associate with potential mismanagement was minimal.

mixture rule (40 CFR 261.3(a)(iv)), derived-from rule (40 CFR 261.3(c)(2)(i)), and policy that the presence, at undefined concentrations, of constituents from listed hazardous waste "contained in" environmental media require the media to be managed as a hazardous waste, materials that are already below characteristically hazardous levels must be managed and treated as hazardous waste amount to punitive acts of repentance: The regulations have no scientific basis, and they can only be rationalized by invoking some moral obligation.[207]

In the history of RCRA from 1980 through 1996, there were only been about 100 successful delisting petitions as reflected in 40 CFR 261 Appendix IX. The delisting (exclusion) process was moved from EPA Headquarters to the EPA regions and states, which could result in even more delays, paperwork and inconsistencies among decisions. Judging from some contemporary (1990s) delisting proposed rulemakings (e.g., FR 61(123), 32746-53 and 32753-65, 25 June 1996), the burden of proof placed upon generators is substantially more rigorous to prove that waste is not hazardous than was placed upon the US EPA originally to prove that the waste was potentially

---

[207] Believe it or not, there is a neo-pagan "earth worshipping" philosophy held in many "environmentalists" circles, which is reflected in a desire to punish industry for its "insults" against the earth. The organized neo-pagan church is the Church of All Worlds (CAW) and it can be found on the Internet. There is also a socialist philosophy of punishing industry and the instruments of capitalism for "insults" against workers. These two philosophies have been the basis for some punitive environmental laws.

hazardous. In the two petitions cited above, it took 9 and 17 months of EPA review for the industry petitions to result in proposed rules, which are subject to about 2 months of public comment before final review and rulemaking can proceed. Obviously, delisting by industry is not a process to be begun on a whim!

The delisting rulemaking concept (40 CFR 261.20, 261.3 and 260.22) should be reserved for eliminating or significantly limiting *an entire industry-wide listing* (e.g., F001) from the lists found in 40 CFR 261.31-.33; not for delisting *individual parcels* of waste identified under the general listing category. In contrast, for *individual parcels of waste or plant-specific waste streams* that happen to meet the operational definition of one of the listed wastes (or are trapped by one of the related rules discussed below), it is outrageous to require a formal rulemaking to eliminate that specific material from the system (without changing the actual listing found in 40 CFR 261).

For example, the industry Dithiocarbamate Task Force took the only option allowed by the EPA to change the actual list of listings and sued the EPA to rollback listings of 28 of 29 carbamate wastes. In Dithiocarbamate Task Force v. EPA (CA DC, 95-1249, 11/1/96), the court correctly ruled that EPA was arbitrary and capricious by not even meeting its stated criteria (40 CFR 261.11(a)(3), see comments above) in the listing.[208] However, I believe the court went too far and erred when it implied that waste that is currently properly managed could not

---

[208] Virtually every existing listing has the same flaw.

be considered as hazardous waste because it is not mismanaged. It is true that listing a waste will not stop accidents from happening, thus, train-accidents are not examples of "mismanagement." Accidental releases can be used as examples of what may happen if mismanagement occurs. Moreover, the court's view implies that as soon as a solid waste comes into compliance with hazardous waste management standards, it is no longer a hazardous waste. According to the Bureau of National Affairs (11-4-96) *"...the court vacated the listing because it is likely [the generator] will continue to use hazardous waste landfills as it did before the listing took effect."* I believe that the EPA was right when it argued *"...that [without the listing] it had no way of knowing if [the generator] would continue to ship its waste to lined landfills."*[209]

Overall, the listing regulations created by EPA under RCRA have created havoc in the identification of hazardous waste and have had consequences in other areas (such as environmental remediation, where lightly contaminated soil or water was forced to be managed as hazardous waste for fear of incurring liability under CERCLA, see below).

Overall, I believe that EPA should have approached "listed waste" as a "rebuttable presumption" (a regulatory concept used in FIFRA regulations), which could be completely rebutted for any specific parcel of material by applying the characteristic test

---

[209] Once again, we see lawyers' minds at play here.

(or showing that the concentrations of listed constituents are below other regulatory levels). *No other paperwork required.*

Also, under RCRA:

Under RCRA, a program for dealing with leaking underground storage tanks (LUST) was initiated in the late 1980s and it required upgrading under-ground storage tanks (e.g., the large gasoline tanks at filling stations) with leak detection, corrosion prevention and spill/over-fill protection.

EPA also had an initiative to mitigate plastics pollution, which was tied to the recycling elements of the program. Some plastics are economically recycled. These plastics are typically (i) expensive to make and (ii) can be feasible depolymerized to a form that can be repolymerized. These are exactly the properties that make aluminum (and most metals) recyclable. No one wants to receive milk in a reused milk bottle unless it has been rigorous cleaned and inspected (glass soft drink bottles were (in to the 1970s) cleaned and reused); but unlike glass, plastic cannot be easily heated and sterilized with high-pressure water. Unlike aluminum or glass, attempting to clean, certify and re-label plastic containers would typically cost much more than making virgin containers. One of the few plastics that can be economically recycled (for a similar use) is polyethylene terephthalate (PET). This polyester can be de-labeled and then digested into a form (polymer to oligomer) that can re-enter the production process. Or it can simply be cleaned and re-extruded. Most thermoplastics (polyolefins) can only be cleaned (not sterilized) and re-extruded as a mixture of polymers into some bulk product (such as parking lot bumper barriers or "plastic wood").

2,4,5-T, PCP, PVC and Dioxins

In Section 8.4, pesticides and herbicides were discussed. 2,4,5-T and pentachlorophenol (PCP) became topics of concern in waste management under RCRA. In the late 1980s, the animal toxicity bioassays in one particular species led to the popular view that 2,3,7,8-TCDD was the most toxic compound known to man. Thus, extreme efforts were made to find it in the environment (at the part per trillion level; most carcinogens are regulated at the part per billion level); and soon it was found in many places. It followed that studies of its destruction in incinerators were pursued. To everyone's surprise, the amount of dioxin that was coming out of incinerators was greater than the amount going in. To the mind of non-scientific bureaucrats, it seemed that dioxin had some unusual stability and could not be destroyed. So, for a period of time, the USEPA worked on the assumption that (i) dioxin was the most toxic material in the world and (ii) could not be destroyed.

It was soon found that burning the plastic polyvinylchloride or the fungicide pentachlorophenol in incinerators produced even larger amounts of dioxin and it was assumed (by EPA) that the mechanism of action involved the same mechanism via 2,4,5-trichlorochlophenol. However, by the mid-1990s, it was becoming apparent that (i) the amount of chlorine in the combustion materials was not the principal driver of dioxin production, (ii) copper played an important role, (iii) even suspended salt in sea breezes was enough to produce dioxin in incinerator ash. Soon, it was made clear that the process of making dioxin was as follows:

High temperature destroys *all organic* compounds. *As the combustion gases cool* (1000°C to 200°C) the molecules of CO_2, H_2O, CO equilibrate to soot (graphitic carbon) if the ratio of carbon is high. Under these conditions HCl (gas) formed from combustion of any halogenated compound is catalyzed by copper (Cu(0)/Cu(I)/Cu(II)) into an electrophilic form that reacts with aromatic hydrocarbons that are formed along the edges of soot particles and dioxins are one of the many forms of relatively stable compounds that are formed. The trick to limiting the formation of dioxins is to quickly quench the exhaust gases through the range (1000°C to 200°C) to minimize the time for dioxins to form. [210]

The presence of traces of catalytic copper are more important to the total amount of dioxin formed than the amount of chlorine. The EPA conducted combustion tests of wood treated with anti-microbial preservatives chromates copper arsenate and pentachlorophenol. Interestingly, the copper-containing wood

---

[210] Waste incinerators were originally designed to maintain a moderate temperature in the flue gases as they passed though the particulate separators and were conducted up tall smoke stacks. This prevented condensation of water and acids that would corrode equipment. Municipal waste incinerators were dioxin manufacturing systems. Now, backyard "burn barrels" are regarded as a major dioxin source.

produces about as much dioxin as the pentacholphenol treated wood, but the pentachlorophenol treated wood gave higher levels of chlorination. Obviously, the dioxins were not being formed by direct reactions of chlorophenol. [211]

Thus, the dioxin scare has waned and to avoid chloracne in manufacturing 2,4-D has largely replaced 2,4,5-T.

Comprehensive Environmental Response, Compensation and Liability Act (CERCLA)

While TSCA and RCRA addressed pressing concerns for new chemical exposures, a different act was needed to deal with existing (legacy) contamination. The example that the USEPA used as their "poster child" to get Congress motivated was an industrial landfill in Niagara, NY known as Love Canal (the history is described above). This partially-dug canal was acquired as a chemical dump and then closed. Subsequently a residential community was built over it and residents discovered chemical odors seeping into their basements. Naturally, they were concerned.

This topic requires some reference to our Constitution. Under our Constitution (1787). Article 1 section 9: *"No Bill of Attainder*

---

[211]Tame NW. et al. 2007. Formation of dioxins and furans during combustion of treated wood *Progress in Energy and Combustion Science.* 33(4):384-408

or ex post facto Law shall be passed."[212] What does this mean? I generally explain it to students this way:

> Suppose a Congress came into power that was opposed to a controversial activity such as "abortion." While the Congress might pass laws making abortion illegal (in the future), the Congress could not declare that doctors that practiced abortion in the past were criminals. By the same token, Congress can pass laws making random or uncontrolled dumping of chemicals illegal *in the future*, but they cannot make people liable for *actions taken legally in the past.*

Thus, when the Congress took up the issue of existing contamination of soil and water that occurred at a time when no law prevented the activity, they had to introduce a mechanism for paying for the remediation of such contamination. They did this by imposing a tax on the importation of manufacture of a number of commodity chemicals and those funds were placed into an account called the *Superfund*. This became the common name of the law that is officially the Comprehensive Environmental Response, Compensation and Liability Act (CERCLA, 1979).

This act prohibited "release" of "hazardous substances" to the "environment." Each of those terms has a specific definition under the Act. That was clearly proactive and was not tricky to deal with. The part that involved the legacy sites was much more complicated.

---

[212] Under section 10, States are also forbidden from passing ex post facto laws.

First off, the law gave EPA authority to go onto *private* property to investigate and remediate contamination…implicitly at *its own expense using the funds* in the Superfund. But the rules that came out of EPA regarding property became a quagmire. The mechanism that EPA created to discover and address contamination was to require that future land transfers (real estate deals) would make the buyer (the new owner) "strictly, jointly and severally" liable for any contamination that existed on the property (*regardless of when or by whom it was created*) **unless the new owner had done *due diligence* to ensure that the property was not contaminated before taking possession. In that case, the new owner was called "the innocent land owner" (see below).**

The terms of liability made the new owner (who had done nothing but buy the property) completely responsible for the contamination.

In practice, the EPA would attempt to identify *potentially responsible parties* (PRPs) and include the "innocent land owner" just as liable as the people who previously owned the land or illegally dumped "hazardous substances" on the land. This became a little confusing, because the actual people who had done the dumping were likely protected by the fact that they did the dumping when it was legal. Thus, the new owner was the person who was most likely to be held liable for the contamination. The new owner might then sue the previous land owner for not disclosing defects in the property when it was sold. Through this chain of events, EPA could potentially ensnare all the previous owners in its liability …making them responsible for investigation and remediation of the contamination. What EPA was looking for was some entity that had enough money to pay for the work, *which could easily cost more than the value of the land.* This was called looking for someone with "deep pockets."

In actual practice, most real estate deals are usually financed by banks. Thus, if a new owner found themselves liable for some exorbitant amount of money (which could easily happen) "they" (i.e., often some land venture corporation) might just declare bankruptcy and the bank would either lose its investment or have to take over the responsibility for the contamination) as the new owner) when they foreclosed on the property. **Thus, the banks (lending institutions) became the parties who forced compliance by demanding "due diligence" be performed before making loans.** There was some question as to what constituted "due diligence" through the 1980s, but as the real estate industry moved into the 1990s, the American Society of Testing and Materials (ASTM)[213] passed a standard for investigation that was feasible, affordable and could be done on a time-scale that was consistent with the speed of real estate transitions.

 There was a requirement under the law, for people who discovered contamination to report it to EPA. Thus, those doing the due diligence ("environmental professionals") were going to report anything they felt was problematic. EPA then could

---

[213] When I first heard that someone was going to have *a committee* develop this standard, I was very skeptical. But I was very impressed with the practicality of these standards and their lack of ambiguity. This erased a major roadblock and financial risk for the real estate industry.

follow up. This could also be a problem for land owners/lenders who had acquired the land after the CERCLA regulations went into effect because they were no longer protected by the *ex post facto* clause.

The potential for litigations was massive and manifest. EPA quickly spend the money in the Superfund and adopted a generally expansive policy during the 1980s. Everybody involved was like to be sued by the EPA. The cost of litigation could easily bankrupt small businesses.[214]

Once it was decided who was going to pay for the investigation and remediation, there were a series of (expensive) steps that were written into to the regulations:

Preliminary Assessment: This was an investigation that would normally be equivalent to what a private sector contractor would call a "phase I site assessment" (i.e., due diligence of the lowest level without any physical investigation other than a "walk over" inspection). Most banks required this step (ASTM protocol) as a condition of loan. It involved records searches and aerial photographs, and if the real estate was "clean" it stopped there.

> As an aside, these were generally done for a couple of thousand dollars and were "lost leaders" for the larger companies providing the service. Phase I site assessments

---

[214] I was an expert witness in several cases in which my client was attempting to force other parties to share liability. There were days of depositions when labor unions, banks and large corporations went at one another. *Hint, when lawyers that make $500/hour talk to one another nothing gets cleaned up; but when a backhoe operator making $25/hr works things get cleaned up.*

were not in themselves profitable, and the companies providing these services were highly incentivized to find something wrong with the property. This is a matter of business ethics and many small companied on the fringes of the industry had no ethics. I have actually seen contractors create fake issues, overstate issues, and argue privately (against the interest of their customer regarding the need for more investigation) with regulatory officials. Even in reputable companies inexperienced and zealous employees frequently erred in the direction of unnecessary investigation.

Regardless, if something were spotted in the records or on the ground, that suggested that contamination by a "hazardous substance" existed, a "phase II" investigation typically would proceed with soil borings, soil gas analysis, soil or surface water sampling, ground penetrating radar, etc. to identify a source of contamination. If this was accomplished, the project shifted to the regulatory authority…in most cases the real estate transaction ended.

EPA had its own contractors who were typically assigned to do the preliminary assessment. If they confirmed the private sector findings, things could get very expensive. At this point, EPA would start looking for *potentially responsible parties* to pay for the work. The process might lapse into a period of discussions, negotiations and litigations among the PRPs and the government agencies (sometimes the federal and sometimes the State and most times both participated). Money starts flying out the window…and the lawyers are the only ones benefiting.

Finally, there is a decision and a full **Site Investigation** (SI) (equivalent to the phase II) approved by the regulators is undertaken. Again, if it confirms contamination, CERCLA requires a new document called a **Remedial Investigation** (RI).

The goal of the RI is to investigate the site in substantial depth to establish the extent of contamination, the movement of contamination, the composition of the contaminants, the likelihood and extent of exposure, etc. The parties have to have a plan for the RI, which can be used as the basis for costing and assigning liability. Just doing the plan for the RI takes time, negotiations, and money. Once everyone (PRPs and governments) is in agreement with the plan, the actual RI generally involves the combined activities of ever sort of environmental scientist and engineer. The studies can easily cost millions of dollars and require many months to complete. If unexpected conditions are encountered, the RI can be extended and re-negotiated. The RI can easily become more of a "science project" than a project targeted at actually remediation of the project. Ultimately, the IR was expected to results in a formal risk assessment.

Under the original CERCLA, remedial design was based entirely on a **Risk Assessment**. The risk assessment included both human risk (usually based on the LNT cancer protocol and standards) and impacts to endangered species. These are the only risk considered in a typical CERCLA analysis.

The RI is then followed by a **Feasibility Study** (FS). The FS is an analysis of remedial options and recommendation of a course of action to mitigate risk. You can mitigate risk two ways: you can remove the contamination or you can separate the endangered people/wildlife from the contamination. In most cases, it was more economical to build a barrier…e.g., pave over contaminated soil. Thus, the early remedial plans (to the extent that the PRPs could control the process) tended to go that way. This trend in mitigating risk greatly upset the sensibilities of many "environmental advocates." Especially, those people (in and out of government) who consider the earth to be sacred (e.g., earth religionists…I'm serious, we are talking about neopagans here…Earth First, the Gaian Hypothesis[215]).

I did not become fully aware of this effect until I was involved in developing a Programmatic Environmental Impact statement for remediation of contaminated sites by the Department of Energy (1991-95). Ultimately, I was accused by a neopagan who had apparently been planted in the contractor team by the Clinton Administration for reprisal against a whistleblower. This was absurd because he was an employee of the prime contractor and I was an employee of a subcontractor. By law, I could not reprise against him because I had no control over him.

The neopagan (who was unexpectedly hired by the prime contractor shortly after the election of the Clinton administration in 1992 with Albert Gore as the Vice President) had complained to me regarding my summary of the use of risk assessment in CERCLA (which he did not know had changed, see below). Many months later, he was fired/laid off by his (prime contractor) management (after he had done his best to redirect

---

[215] Recall James Lovelock and the GC-ECD. In practice, this hypothesis elevated "mother nature" (a pagan religious concept) to the status of God and regards any manmade blemish to the earth as a sin. Thus, leaving contamination in place is a sin.

the program to his liking). He then brought a whistleblower complaint against his employer, but he claimed that his firing was a reprisal because of me…I did not hire him, I did not review his work, I never complained to his management about his behavior, and I did not even know when he was fired. But that did not stop the Department of Energy (Clinton Administration officials) from highlighting me as the guilty party in the whistleblower complaint proceeding (which I was never asked to participate in and did not even know occurred until after the fact).

Yes, this is a true story. Long after I was separated from the subcontractor who I worked for, my name appeared on the Internet (1996) as the person who had reprised against this guy, I was dumbfounded and worried. I knew that the Internet would be used to do due diligence by future employers. I started researching him and discovered that he was a neopagan. Literally, a self-proclaimed practicing witch[216]:

A typical example of his (Larry Cornett's) scientific understanding is provided below in part. As you read this remember that this man won a whistleblower complaint against his employer (the prime contractor) and had the US Department of Energy libel me on the Internet although I did not even work for his company (I worked for a subcontractor):

216

https://www.paganlibrary.com/reference/nature_spirit_magic.php

Nature Spirit Magic

by Larry Cornett

https://www.paganlibrary.com/reference/nature_spirit_magic.php

Introduction

Each plant, animal, rock, and other entity has a spirit (consciousness resonance matrix). These spirits can join together, in a hive-mind, as a spirit of an area. Nature spirits include real biological intelligences, are psychically powerful, and are much less abstract and controllable than the Elementals that many magical people who perform all of their rituals indoors are familiar with. They can be extremely powerful allies. It is possible to sense nature spirits, to determine if they are receptive to a ritual planned, and to have them actively participate in magical workings if they are.

Some Effects Of Working With Nature Spirits

Spectacular physical manifestations can happen when working with nature spirits in the wild. I have personally seen actual foxfire mark the boundaries of a magic circle at a location that was identified as a receptive power spot and attuned to a planned ritual the day before. I have seen more than one site attuned for ritual be dry and comfortable, with a round hole in the clouds overhead, on days that were cold and rainy at other nearby locations. Birds have joined in rituals, flying around the circle when energy was being raised; and insects, birds and animals have joined in chants. In addition, the wind often responds to invocations. Generally, these spectacular manifestations happen unexpectedly.

With or without such manifestations, nature spirits often will channel tremendous amounts of power into the magic being performed. It is suggested that you do not consciously try for specific manifestations. Let Nature channel her power into the magic in her own way. If approached with respect, nature may give you many pleasant surprises.

Spectacular physical manifestations are not a necessary sign of success. If you need a spectacular manifestation and nature spirits know this, you will get it. The best success in magic is on the inner planes and more subtle than such manifestations. This success involves beneficial changes in consciousness that last and helpful chains of synchronicity. In addition, working with Nature

Spirits can also bring a deep sense of partnership with Nature, and bring new levels of attunement.

To get the best results, perform nature spirit attunement several hours to several days before the main ritual. The purposes of such attunement are to find suitable power spots and to get the help of friendly nature spirits. This timing gives Nature time to gather her children and to prepare to actively participate in the main ritual.

What To Not Do

If nature spirits are approached with disrespect by attempting to command them rather than listening to them and inviting them to work with you, nature spirits may flee, rebel, or attack. I once attended a ritual by some pseudo-Crowleyites who attempted to perform the "Ritual of the Barbarous Names" at a power spot in a forest and then to extend the circle several hundred yards in all directions.

While the forest in general had loud insect and frog noises, the area at which the ritual took place got quiet immediately when the main ritualist declared that all spirits were subject unto him. The vibes from nature could best be characterized as "Oh yea, Mother!" One participant was quickly possessed by an angry spirit and kept repeating "You killed my children, your children will never live in peace." When the priestess stepped out of the boundaries of the original circle, she was attacked by bees; and bees covered the Book of the Law. Magicians should know better than to attempt to command spirits whose true names they do not know!

Calling Nature Spirits

To make the most out of working magical ritual in the wild, one should find power spots where nature spirits are receptive to the ritual planned and approach the spirits with respect, as equals. In my experience, the most effective power spots for working with the living intelligences of nature are located in wild areas with diverse, active ecologies.

AND SO FORTH...

The issue that precipitated this retribution by Cornett (and his Clinton Administration allies[217] at the DoE) against me was a document I wrote as part of the PEIS team in which I pointed out that the role of risk assessment in CERCLA was vastly changed by the *Superfund Amendments and Reauthorization Act* (SARA) of October 17, 1986.

Recall the original CERCLA was implements in 1980 and included an elaborate rationale for risk assessment, which was based on the same criteria (target 1-in-one-million *lifetime risk*) discussed above. Because many sites were being remediated by "control" strategies rather than "clean-up" strategies, many people were discontent with the idea of minimizing risk as the driver of a Feasibility Study that resulted in a government-approved **Remedial Plan**. Thus, SARA included a requirement to achieve all **Applicable or Relevant and Appropriate Requirements** (ARAR) as a condition of successful remediation.

With regard to the whistleblower's charge, after the passage of SARA, most remedial plans were driven by the requirement to achieve ARAR

---

[217] The PEIS was being written because of a law suit by professional environmental advocates. The Clinton administration hired those advocates into the Department of Energy to oversee the PEIS and related programs. I suspect that Cornett was hired by the prime contractor at the direction of these people for the purpose of influencing the outcome of the analysis…in my opinion, he had no obvious qualifications for the job, but was initially given great deference by the project management (e.g., I only briefed two new hires of the prime contractor…the new manager and Larry Cornett). I was likely targeted because I had been the leading briefer of the Assistant Secretary for the Environment on two occasions.

standards rather than by the risk assessment process. Indeed, achieving ARARs could force clean-up even in the absence of any risk.

I do not know, but I suspect that the company ICF, which was a very influential EPA contractor (with political connections) had a large hand in guiding EPA down this path. There were rumors that there were "Champaign parties" at ICF when this legislation past, because they knew that it would require a much more expensive approach to remediation. About this time, ICF (which had been primarily a policy shop) acquired Kaiser Engineering to make a company called ICF-Kaiser (1988) with the intension of moving beyond consulting into environmental remediation (engineering and construction work were where the really big money was) and they "went public" (i.e., became a publicly owned company trading on the stock market, which typically makes the founders rich).[218]

For the ARAR idea to maximize clean-up (and maximize the profits in companies engaged in that work), the EPA policy was to regard all regulations that even vaguely applied to a site of contamination to be enforceable in the drafting on the Feasibility Study and Remedial Plan.

The most common example of ARAR invoked in the Feasibility Study was the drinking water MCLs under the Safe Drinking Water Act. These standards were easy to identify and the position of most regulators (who tended not to know the details

---

[218] ICF has been a driving force at EPA influencing regulatory policy behind a number of federal environmental programs (including the ozone, global warming/climate change, and renewable energy initiatives). The Kaiser activity did not work out well for them. By 1999, the consulting division jettisoned Kaiser.

of the SDWA) was to enforce them on every drop of contaminated water as though they were "applicable." But in most cases, the SDWS MCLs were clearly not "Applicable" indeed by using the acronym EPA managed to obscure the fact that there was actually a legal argument to be made here. First off, if no one were drinking the water, the MCLs were not *applicable*. Were they *relevant and appropriate*? I would argue from my experience in the field that the answer is "No." If you look at the SDWA (no one working on CERCLA did) you find that private wells are not regulated subject to the MCLs and indeed the water systems that are regulated have to be of a relevant size (i.e., they must produce enough volume of water over a long period of time) and the source of water should implicitly be free of natural contaminants (including bacteria common on the surface). Nonetheless, the MCLs were generally used to drive remediation of small pockets of near-surface "perched water" (e.g., a subsurface "pond" of water perched in a permeable layer (of sand) on an impermeable layer (of clay). These near-surface pools were often contaminated by downward percolation of liquid compounds. But they did not represent enough volume to ever be useful as a drinking water supply regulated under SWDA and basic construction of water supply wells would never accept a pool of water a few feet below the surface (subject to direct infiltration of bacteria) as a water source for a small community. The well suppling my house is over 100 feet deep and is sealed from the surface to the aquifer by impermeable barriers both natural and associated with the well itself. Obviously, I have a septic tank and infiltration field in my front yard a few feet under the ground and the well is in the back yard (100 feet away) but much deeper.

In any event, eventually the site gets cleaned up.

One of the major problems faced in the 1990s (I think it has been resolved over the last 20 years) has to do with **investigation derived waste** (IDW). This is actually a problem caused by the listed waste "contained in" rule under RCRA and the "strict, joint and several liability" standard under CERCLA. If lawyers get involved all hell breaks loose. On the one hand, if soil or water is contaminated with a "listed waste," under the RCRA contained-in rule, the entire parcel of contaminated soil legally becomes a hazardous waste. CERCLA forbids release of "hazardous substances" (which includes "hazardous waste") to the environment (after 1980). Thus when a contractor digs up contaminated soil (even if it is very lightly contaminated with a "listed waste") and places it on the ground, they just became a potentially responsible party and are severally responsible for the ENTIRE cost of clean-up of the entire site.[219] In the aggressive enforcement and punitive mind-set of EPA officials in Washington and in the field, this threat was taken very seriously by any company that had substantial assets.

In about 1992 (concurrent with my work on the DOE PEIS), I became the project manager for an installation-wide site

---

[219] Business Dictionary: *"US term for the liability for cleanup of hazardous wastes. Here 'strict' means that wrongdoers are liable whether or not they acted carelessly or unreasonably; 'joint and several' means that any and all of the wrongdoers can be forced to pay for all the damages in case of an indivisible harm (where the proportion of total harm attributable to each wrongdoer cannot be determined)."*

Read more: http://www.businessdictionary.com/definition/strict-joint-and-several-liability.html

assessment of Fort Riley, Kansas. When we got there taking over from a different contractor, I discovered that there were about 800 55-gallon drums of "hazardous waste" sitting on the flood plain of the Kansas River between Camp Funston and Main Post. These drums contained soil, water, various types of debris even protective clothing and pieces of concrete that the precious contractor had (in the opinion of their experts) declared as hazardous waste and had dutifully containerized. One of my jobs as program manager (and having been involved as a consultant to the RCRA "listing program" of EPA) was to develop a safety plan for work on the site. I wrote the plan to require upgrading to "level B" protective masks and clothing if (and only if) field monitors detected hazardous levels of contamination. In turn, my "IDW management plan" proscribed that (IDW) waste need not be containerized unless this safety upgrade were required. In other words, if the soil and water we were handling did not cause us to have to upgrade our protection for field workers, *we had no reason to believe* that the soil or water was contaminated (much less contaminated with a "listed waste" as opposed to a "characteristic waste"). Thus, over the 5 years we were at the site, we never generated but one drum of waste (almost pure gasoline pumped from a well in an area of a recent spill).

As a funny story, at one point, I went to a meeting at Fort Riley in the middle of a great rain storm (May 1995?). When I got there the river had flooded and I was told to return to Washington, DC because my client was in a helicopter trying to track down the 800 drums marked as "Hazardous Waste: Fort Riley." These drums were floating down the Kansas River toward Topeka, the state capitol.[220]   After that we were asked to bid on the disposal of the waste. We made the assumption that the waste would merely be solid waste and could be disposed on one of the on-site landfills. Even with that, the testing and work that would be required to prove that the contents of the drums were not "characteristically hazardous" and the labor needed to treat it as hazardous waste pushed the cost of the project into the millions. The deal was never struck, but by the next time I visited my client, the contents of the drums had been dumped in a construction land fill and the steel drums had been crushed for recycle. I suspect the commanding general got involved.

---

[220] Perhaps you cannot imagine the embarrassment that this would have caused the Army if a picture of one of these containers washed up onto a beach of the Kansas River got into the national press. I can.

www.ingramcontent.com/pod-product-compliance
Lightning Source LLC
Chambersburg PA
CBHW030604220526
45463CB00004B/1164